T0192872

Steel Corrosion and Degradation of its Mechanical Properties

Steel Corrosion and Degradation of its Mechanical Properties

Chun-Qing Li and Wei Yang

CRC Press is an imprint of the
Taylor & Francis Group, an **informa** business
A BALKEMA BOOK

CRC Press/Balkema is an imprint of the Taylor & Francis Group, an informa business

Typeset by codeMantra

Library of Congress Cataloging-in-Publication Data Applied for

Published by: CRC Press/Balkema
Schipholweg 107C, 2316 XC Leiden, The Netherlands
e-mail: Pub.NL@taylorandfrancis.com
www.crcpress.com – www.taylorandfrancis.com

ISBN: 978-0-367-63586-2 (hbk)
ISBN: 978-1-003-11979-1 (eBook)
ISBN: 978-0-367-63590-9 (pbk)

DOI: 10.1201/9781003119791

The book is dedicated to my parents, Li Guanqun and Liu Zhengyuan, who endured all the suffering of hard life, but did not survive to enjoy comforting of better life.

– Li Chunqing

Contents

Preface xi
Acknowledgements xv

1 Introduction 1
1.1 Background 1
1.1.1 Brief history of steel 1
1.1.2 Advantages of steel 3
1.1.3 Application of steel 4
1.2 Significance of the book 6
1.2.1 Corrosion-induced failures 6
1.2.2 Corrosion-induced costs 8
1.2.3 Current practice 9
1.3 Purposes of the book 10

2 Basics of steel corrosion 13
2.1 Introduction 13
2.1.1 Making of steel 13
2.1.2 Chemical composition of steel 14
2.1.3 Mechanical properties of steel 16
2.2 Corrosion process of steel 17
2.2.1 Electrochemical reactions 17
2.2.2 Progress of corrosion 19
2.2.3 Types of corrosion 21
2.2.3.1 Uniform corrosion 22
2.2.3.2 Pitting corrosion 22
2.2.3.3 Crevice corrosion 23
2.2.3.4 Microbial corrosion 24
2.3 Factors affecting corrosion 24
2.3.1 Environmental factors 24
2.3.1.1 Concentration of dissolved oxygen 24
2.3.1.2 Temperature 25
2.3.1.3 Relative humidity 25
2.3.1.4 pH value 26
2.3.1.5 Salts 26

2.3.2 Material factors 27
2.3.2.1 Chemical composition 27
2.3.2.2 Microstructure 27
2.3.2.3 Defects 28
2.3.3 Soil factors 28
2.3.3.1 Water content 28
2.3.3.2 Soil resistivity 28
2.3.3.3 Soil pH 29
2.3.3.4 Soil texture 29
2.3.3.5 Other factors in soil 30
2.4 Effects of steel corrosion *30*
2.4.1 Physical effect 31
2.4.2 Chemical effect 32
2.4.3 Microstructural effect 33
2.5 Corrosion characteristics of ferrous metals *34*
2.5.1 Difference in material 35
2.5.2 Difference in corrosion 36
2.5.3 Comparison of corrosion 36
2.6 Summary *39*

3 Corrosion impact on mechanical properties of steel **41**
3.1 Introduction *41*
3.2 Observation of corrosion *42*
3.2.1 Simulated corrosion 42
3.2.1.1 Exposure environments 43
3.2.1.2 Test specimens 44
3.2.1.3 Test procedure 45
3.2.2 Natural corrosion 46
3.2.3 Corrosion measurement 47
3.3 Degradation of tensile properties of steel *53*
3.3.1 Reduction of yield strength 55
3.3.2 Reduction of ultimate strength 60
3.3.3 Reduction of failure strain 64
3.4 Degradation of fatigue and toughness properties of steel *66*
3.4.1 Reduction of fatigue strength 66
3.4.2 Reduction of fracture toughness 70
3.4.3 Comparison of mechanical properties 75
3.5 Mechanism for degradation *77*
3.5.1 Changes in element composition 78
3.5.2 Changes in grain size 84
3.5.3 Changes in iron phase 85
3.6 Summary *87*

4 Corrosion impact on mechanical properties of cast iron and ductile iron **89**

4.1 Introduction *89*
4.2 Observation of corrosion of cast iron *90*
4.2.1 Simulated corrosion 90
4.2.1.1 Exposure environment 91
4.2.1.2 Test specimens 92
4.2.1.3 Test procedure 94
4.2.2 Natural corrosion 95
4.2.3 Corrosion measurement 96
4.3 Degradation of mechanical properties of cast iron *104*
4.3.1 Reduction of tensile strength 104
4.3.2 Reduction of modulus of rupture 107
4.3.3 Reduction of fracture toughness 110
4.4 Degradation of mechanical properties of ductile iron *119*
4.4.1 Observation of corrosion 120
4.4.2 Reduction of fracture toughness 120
4.4.3 Comparison of mechanical properties 122
4.5 Mechanism for degradation *124*
4.5.1 Changes in element composition 124
4.5.2 Changes in iron phase 129
4.5.3 Pitting corrosion 132
4.6 Summary *133*

5 Other corrosion damages **135**

5.1 Introduction *135*
5.2 Stress effect on corrosion *136*
5.2.1 Observation of stress effect 137
5.2.2 Effect on microstructure 140
5.2.3 Effect on mechanical properties 143
5.3 Preferred corrosion *146*
5.3.1 Causes of preferred corrosion 147
5.3.2 Factors affecting preferred corrosion 149
5.3.2.1 Solidification speed 149
5.3.2.2 Elemental composition of steel 151
5.3.2.3 Temperature of steel casting 151
5.3.3 Prevention of preferred corrosion 152
5.4 Corrosion-induced delamination *153*
5.4.1 Observation of delamination 154
5.4.2 Quantification of delamination 155
5.4.3 Mechanism for delamination 156
5.5 Hydrogen embrittlement *160*
5.5.1 Observation of hydrogen concentration 162
5.5.2 Effect of hydrogen concentration 164

5.5.3 Mechanism for hydrogen embrittlement 165
5.6 Summary *169*

6 Practical application and future outlook 171
6.1 Introduction *171*
6.2 Calibration of simulated tests *172*
6.2.1 Basics of similarity theory 172
6.2.2 Acceleration factor 173
6.2.3 Examples 176
6.3 Practical applications *178*
6.3.1 General procedure 178
6.3.2 Steel structures 180
6.3.3 Cast iron structures 184
6.4 Simultaneous corrosion and service loads *187*
6.4.1 Testing methodology 188
6.4.2 Combined corrosion and bending 189
6.4.3 Combined corrosion and fatigue 191
6.5 Nanomechanics of corrosion *193*
6.5.1 Basic idea 193
6.5.2 Mapping of atomic lattice 195
6.5.3 Model development 197
6.6 Summary *199*

Bibliography 201
Index 221

Preface

There have been many books on steel corrosion, and yet another one is coming. Why? Because the book is not really on steel corrosion itself from an electrochemical perspective, but it focuses on corrosion effect on degradation of steel from a mechanistic perspective, and it is not just on steel but also on ferrous metals in general with examples of practical applications. There are of course many compelling reasons for the need of this book, of which the most important one is that steel corrosion continues to not only cause damages to steel structures, the costs of which run in billions, but also and more importantly lead to ultimate collapses of steel structures, the consequences of which are catastrophic and beyond estimation. The reoccurrence of corrosion-induced structural failures, however defined, exposes the inadequacy of current knowledge on steel corrosion and in particular, its effect on degradation of mechanical properties of steel. This taxes the capability of researchers and engineers to predict and prevent failures of corrosion affected/prone steel structures. It suggests that there is a gap in current knowledge on steel corrosion, which is its effect on degradation of mechanical properties of the corroded steel. Without this knowledge, structural collapse due to steel corrosion can hardly be predicted as it is the mechanical properties of the steel material that provide strength to the structure to prevent its collapse. This gap in knowledge hinders further advancement of corrosion science and hampers its application to accurate prediction of collapse of corroded steel structures. It is in this regard that the book is solicited.

Although corrosion of steel is a well-trodden topic with a long history, there has been so much research conducted on this topic, and yet, there is still so much unknown and hence so much to explore. The fact is that steel continues to corrode, and steel structures continue to collapse, as evidenced in the book. It would be immature to think that one book has covered or can cover all the knowledge of corrosion both from a science perspective, such as electrochemical reactions of corrosion, and an engineering perspective, such as corrosion effect on mechanical properties of steel, in particular the prediction of degradation of mechanical properties of corroded steel. Whilst it is fully acknowledged that corrosion science lays the foundation of corrosion engineering, it is the corrosion effect on degradation of mechanical properties that matters in the real world of steel and steel structures. How to accurately predict the corrosion effect on the degradation of mechanical properties of the corroded steel is and will continue to remain a serious challenge to all stakeholders in corrosion research and practice.

The aim of this book is to present the state-of-the-art-knowledge on corrosion of steel, cast iron and ductile iron, collectively referred to as steel in short. It focuses on corrosion-induced degradation of the mechanical properties of corroded steel, which is examined at both macrolevel, e.g., the mechanical strength, and microlevel, e.g., element composition, and also with and without service loading. Knowledge on corrosion-induced degradation of mechanical properties of corroded steel is largely unavailable but very much needed to accurately assess and predict failures of steel structures due to corrosion. This knowledge is imperative in the planning for maintenance and repairs of corrosion-affected structures. The information presented in the book is largely produced from the most current research on corrosion effect on degradation of the mechanical properties of steel. The book covers the basics of steel corrosion, including that of cast iron and ductile iron that are not well covered in most literature. Models for corrosion-induced degradation of mechanical properties of steel are also presented as much as possible in the book with a view of wider practical applications. These models can equip practitioners in accurately assessing and predicting corrosion-induced failures of steel and steel structures. As is known, the cost related to corrosion-induced maintenance, repairs and unexpected collapses is beyond estimation. Therefore, the knowledge presented in the book can be used to prevent corrosion-induced failures, producing huge benefits to the industry, business, society and community.

This book takes a quite different approach in dealing with corrosion of steel, which is more mechanistic, focusing on physical process, more phenomenological, focusing on observations, and more practical, focusing on end results for practical applications of the developed knowledge, and yet without losing insightful analysis of fundamental causes of corrosion and its effect on degradation of mechanical properties of corroded steel. The significance of the book is that it presents much needed knowledge on corrosion-induced degradation of mechanical properties of steel, which can be used for service life prediction of corrosion affected/prone structures made of steel, cast iron and ductile iron, preventing their failures and planning for maintenance and repairs for such structures.

A few characteristics of the book stand out. The first one is its usefulness. Corrosion is a global problem faced by various researchers, professionals and the public at large. This book can be employed by a variety of users, such as students/trainees who attempt to learn more about corrosion and its effects, academics/researchers who like to teach and research in this area, practitioners/engineers who try to design and assess corrosion affected/prone structures and importantly, asset managers/owners who need to ensure the safety and serviceability of corrosion affected structures. The second characteristic is practicableness. This book aims for practical applications with many real-world examples. The knowledge and techniques presented in the book can be easily applied by users, such as material and structural engineers who are responsible for safe and serviceable operation of corroded structures, asset managers and owners who need to make decisions regarding the maintenance and even demolition of corroded structures. The third and perhaps the most important characteristic of the book is its uniqueness. This book can be the first of its kind, not just by its title but by its contents, that examines corrosion and its effect on degradation of various mechanical properties, from common tensile strength to more complicated fatigue and most difficult fracture toughness, of various ferrous metals, from commonly known steel to less familiar cost iron and ductile iron. The knowledge on degradation of mechanical properties of corroded steel is vital to accurately predicting and effectively preventing failures of corrosion-affected structures. Last but not least is its readability. The book is written in plain English as much as practicable. Technical terminology and complex electrochemical reactions are either explained in simple terms as much as practical with necessary figures for easy understanding or referred to other specialist books on corrosion for details.

Given the inevitability of corrosion, it is time to consider a paradigm shift in the body of knowledge of corrosion from its diagnosis and protection to prediction and prevention of its damages and consequences. With this thought in mind, we are delighted to present this book to readers.

Chun-Qing Li and Wei Yang
Parkville, Melbourne
Australia
January 2021

Acknowledgements

There is a long list of people and organisations to thank and acknowledge. First and foremost, I am in deep debt to my dearest wife who gave me a new life: encouraging me to keep moving forward, supporting me in difficult times and stimulating me for new ideas. She also helped and involved directly in writing the book.

This book is not possible without the funding from the Australian Research Council (Discovery Projects) and Metro Trains Melbourne, which enabled the realisation of original research ideas and proposals and turned them into research outputs, mostly journal publications. This book is not possible without the contribution of a number of my research fellows, in particular Drs Gaoyang Fu, Hassan Baji and Wenhai Shi, and my PhD students, in particular Drs Muhammad Wasim, Le Li and Weigang Wang, who directly worked on experiments on corrosion of the funded projects, which provide raw data for the book. Their hard work and commitment to the projects in particular experiments are greatly appreciated and thankfully acknowledged. More recently in the process of writing this book, the great help and contribution from Drs Le Li and Weigang Wang in processing test data and figures are thankfully acknowledged. Both Drs Li and Wang have actually drafted some paragraphs of the book, which is highly appreciated and thankfully acknowledged. Some of my current PhD students Yanlin Wang, Bohua Zhang and Yurue Li also contributed to the book by providing reviews of related literature and compiling figures. This is gratefully appreciated. This book is not possible without the contribution of other colleagues who are part of the research projects or team on corrosion, intermittently, in particular, Prof Sujeeva Setunge, Associate Professor Annan Zhou, Drs Dilan Roberts and Mojtaba Mahmoodian, whose commitment to the research projects, inputs to the research activities and support of research team are greatly appreciated and gratefully acknowledged. There are also unsung heroes, in particular laboratory technicians, who contributed to the book by ensuring the successful experiments and quality results.

Last but not least, our newborn baby, Shelin, who obviously could not contribute materially but has contributed spiritually just in time before the book was finished, which is equally precious with more joy.

Chapter 1

Introduction

1.1 BACKGROUND

Corrosion of steel is a well-trodden topic, and yet, it is still causing considerable problems to all stakeholders from manufacturers to designers and end users. There are many books written on steel corrosion, yet there is still a lack of sufficient knowledge on it. What would be new from this book that will keep its place in the ocean of literature on steel corrosion is not the corrosion itself which is supposed to know, but its effect on steel, which is yet to develop. This book discusses more about the mechanical properties of steel that are closely associated with corrosion, which makes it stand out from other books on steel corrosion. Steel in this book generally refers to ferrous metals, including steel, cast iron and ductile iron. In general, or in short, they are collectively called steel in this book.

1.1.1 Brief history of steel

The evolution of humankind is closely related to the use of metals. To some extent, the civilisation of humanities has by and large relied on the discovery and development of metals. As one of the most used metals, iron was discovered by accident when some ore was dropped into a fire and cooled into wrought iron. The first smelting of iron was around 3000 BC which led to the start of the Iron Age (1200 BC). The transition from the Bronze Age to the Iron Age occurred at different times in different places in the world, but when and where it did, the distinctive dark metal brought with it significant changes to daily life in ancient societies, from the way people grew crops to the way they fought wars (www.wikipedia.org).

Iron has remained an essential metal for more than 3,000 years, through the Industrial Revolution and into today in its more sophisticated form, i.e., steel. Even in the modern times, however, the quality of iron produced depends as much on the ore available as on the production methods. By the 17th century, the properties of iron were well understood, but increasing urbanisation in Europe demanded a more versatile metal for structural use. By the 19th century, the amount of iron being consumed due to railroad expansion provided metallurgists with the financial incentive to find a solution to iron's brittleness and inefficient production processes.

The development of blast furnaces increased the production of cast iron. Various methods for reducing the carbon content to make iron more workable were experimented by metallurgists. By the late 1700s, ironmakers learned how to transform cast pig iron into a low-carbon content wrought iron (about 0.1% carbon content) using puddling furnaces. As the carbon content decreases, the melting point of iron increases so that the masses of iron would agglomerate in the furnace. These masses would be removed and worked with a forge hammer before being rolled into sheets or rails.

Blister steel is one of the earliest forms of steel, which began being produced in Germany and England in the 17th century and was produced by increasing the carbon content in molten pig iron using a process known as cementation. In this process, bars of wrought iron were layered with powdered charcoal in stone boxes and heated. Blister steel production grew in the 1740s when English clockmaker Benjamin Huntsman, whilst trying to develop high-quality steel for his clock springs, found that the metal could be melted in clay crucibles and refined with a special flux to remove slag that the cementation process left behind. The result was a crucible, or cast, steel. But due to the cost of production, both blister and cast steel were only ever used in speciality applications. As a result, cast iron made in puddling furnaces remained the primary structural metal in industrialising Britain during most of the 1800s.

The situation with steel being an unproven and yet costly structural metal changed in 1856 when Sir Henry Bessemer, an English engineer and inventor, developed a more effective way to introduce oxygen into molten iron for reducing the carbon content in the iron. Bessemer designed a pear-shaped receptacle, referred to as a "converter", in which iron could be heated whilst oxygen could be blown through the molten metal. As oxygen passed through the molten metal, it would react with the carbon, releasing carbon dioxide and producing a purer iron. The process was fast and inexpensive, removing carbon and silicon from iron in a matter of minutes, but since it was too efficient, too much carbon was removed with too much oxygen remaining in the final product. Bessemer ultimately had to repay his investors until he could find a method to increase the carbon content and remove the unwanted oxygen. Nevertheless, the development of what is known now as the Bessemer process is considered to be the beginning of the modern steel industry.

At about the same time, British metallurgist Robert Mushet started to test a compound made up of iron, carbon and manganese, known as spiegeleisen. Manganese was known to remove oxygen from molten iron and the carbon content in the spiegeleisen. If the correct amount of manganese was added in this compound, it would solve the problem of too much oxygen encountered by Bessemer. The addition of the manganese to the conversion process of iron was a great success. But it was until 1876 when Welshman Sidney Gilchrist Thomas developed an innovative solution to the Bessemer process, i.e., adding limestone, iron ore from anywhere in the world could be used to make steel. As a result, the cost of steel production began to decrease significantly by about 80% between 1867 and 1884, which initiated the growth of the world steel industry. Since then, there were other new developments in the history of steel making, notably the open hearth process, the electric arc furnace and the oxygen furnaces.

Fast forward to the 21st century, the modern process of steel making is all standardised with more advanced technology. As a result, the quality and versatility of steel increase, whilst the cost of steel making decreases. According to the World Steel Association (2019), the crude steel production stands at 1,808 million tonnes in 2018 with an increase of about ten times from 189 million tonnes in 1950.

1.1.2 Advantages of steel

Steel remains one of the most important conduction and machinery materials since its advent. This is because it has many advantages as compared with other building materials, such as concrete, masonry and timber. The most recognised advantages of steel can be summarised as follows (NIST NCSTAR 1-3D 2005):

1. High strength. With the modern technology of steel making, the tensile strength of steel can reach 2000 MPa (N/mm^2). This is incomparable with almost all other building materials. For commonly used structural steel, the strength is in the range of 250–450 MPa which is still much stronger than most building materials. The next strongest building material could be concrete in compression with compressive strength in the range of 30–50 MPa.
2. Great lightness. Steel has the best strength to weight ratio amongst almost all other major building materials. The strength to weight ratio of steel is in a range of 30–40 for commonly used structural steels, i.e., mild steel. This is about three times higher than concrete which is another most used building material. The strength to weight ratio of some high-strength steel, e.g., chromoly steel, can reach 85.
3. Excellent ductility. Steel, in particular, mild steel, has excellent ductility which is its ability to be drawn or plastically deformed without fracture or rupture. Because of good ductility, steel is extremely flexible with many product forms and shapes, easy to bend and tough to fracture.
4. Extreme tautness. Steel can be stretched or pulled tight without slacks. This is because of its good elastic behaviour. Together with other mechanical properties, good tautness makes steel more workable, formable and versatile. Structural steel sections can be bent and rolled to create non-linear members to enhance the aesthetic appeal of the structure. This is in particular useful for steel fabrication in situ or ex situ.
5. Transparency and elegancy. Steel products are either hot-rolled or cold-formed with various cross sections, such as, I section, hollow section, channel section and angles. All of these sections are transparent. With the great lightness, steel members are relatively slender and look more elegant when they are erected. The structures constructed of steel are usually in the form of frame which looks spacious and transparent. Architects admire the natural beauty of steel and emphasise the grace, slenderness, strength and transparency in their designs.

In addition to these advantages, structural steel brings numerous benefits to construction projects, including the speed of construction due to its shop fabrication and site erection resulting in lower project time and costs, aesthetic appeal due to its transparency and slenderness, high strength due to its restraint power, sustainability due to its recyclability, modifiability due to its easy fabrication, innovativeness due to the versatility of steel and its products, efficiency due to optimal use of building space and reliability and predictability due to the high quality assurance processes in steel making. These advantages in steel construction warrant another book to fully describe and appreciate.

Whilst every coin has two sides, steel is no exception, but the disadvantages of steel are not as many as its advantages. The most notable disadvantages are its susceptibility to fire and corrosion since both of these can be catastrophic. The mechanical properties of steel are usually constant with temperature only when it is within a certain limit, depending on the type of steel. For some structural steel, the yield strength and elastic modulus start to decrease when the temperature exceeds 200°C (www.engineeringtoolbox.com). There are many examples of collapses of steel structures under the elevated temperature, i.e., fire. One of the earliest disasters under fire could be the Crystal Palace, which was originally built in Hyde Park, London, to house the Great Exhibition of 1851 and then relocated to an area of South London in 1854. The Crystal Palace was constructed of cast iron and plate glass and completely destroyed by fire in November 1936. The most recent disaster could be the collapse of twin towers of World Trade Center in New York. The structure of both towers was steel tube frame. Under the fire, both towers were destroyed within a couple of hours. Obviously, fire is one of the most catastrophic hazards to steel structures. Another one is corrosion which is the subject of this book.

With the advancement of theories of mechanics and structures and the development of new technology in structural design, fire hazard as a failure mechanism in structural design can be designed out in most cases or can be prevented to some extent. One failure mechanism that may not be designed out is the corrosion of steel with surrounding environments since it is natural. Corrosion-induced failures occur almost every year and everywhere from small mishaps to catastrophic disasters, some of which are to be presented in Section 1.2.1.

1.1.3 Application of steel

Throughout the history of humanities, metals are used for various purposes. Gold is perhaps the first and most valuable metal to be used, which was fashioned into jewellery in the Stone Age (6000 BC). Due to its scarcity, gold was used as exchange of values and throughout the history as one of the bases of monetary values. Copper, on the other hand, was possibly the first metal to be used for practical purposes. Even in the Stone Age, copper was used to make tools, implements and weapons.

It can be said in general that iron and steel can be used anywhere and almost for everything. Undoubtedly, the major use of steel is in construction with 43% of total steel production in 2018 (Li et al. 2018). Steel has been used in almost all structures, including concrete structures, where reinforcing steel is essential to overcome the weakness of concrete in tension, masonry structures and even timber structures where steel is important for, e.g., connection and reinforcement. Historically, the use of iron and steel can be highlighted in some landmark structures in the world.

Cast iron made its debut in the construction industry in 1779 with the bridge at Coalbrookdale in England, which was considered to be the first large-scale use of cast iron for structural purposes. Iron structures soon began to find their way into textile factories, stage construction and glasshouses. The central market building in Paris, Les Halles, built in 1853, was the first building in France to openly display its metalwork. It opened the way for the construction of new types of edifice required by an industrialised society, such as railway stations, markets, factories, large stores, glass-roofed buildings, pavilions and exhibition halls. The use of iron in architecture spread widely and became one of the most original and spectacular forms of creative expression of the nineteenth century because of its lightness, its transparency and the elegant way it rises into the air, coupled with its brute strength, its restrained power and its extreme tautness.

Amongst these many magnificent structures made of iron and steel, the Eiffel Tower (Figure 1.1) excels all and has become a global cultural icon of France and one of the most recognisable structures in the world. It is also a monumental example of material properties and structural performance. The Eiffel Tower was constructed of wrought iron from 1887 to 1889 as the entrance to the 1889 World's Fair. A total of 7,000 metric tons of iron was used. The tower is 324m tall, about the same height as an 81-storey building and remains the tallest structure in Paris. It was the first structure to reach a height of 300m. By the year 1885, the time when the Tower was being constructed, the use of iron and steel in bridges and building frameworks had become widespread.

The bridge over the Firth of River Forth in Scotland is another such example. Dubbed as Scotland's Eiffel Tower, the breath-taking Forth Railway Bridge stands at Queensferry Narrows, about 15km west of Edinburgh, where it carries trains for one and a half miles over the Firth of River Forth. The Forth Railway Bridge is a

Figure 1.1 Eiffel Tower.

remarkable cantilever structure which is still regarded as an engineering marvel. The structure of the bridge consists of its three massive cantilever towers each 104 m high and achieves a record span of 521 m. The construction began in 1883, and after 7 years, with 55,000 tons of steel, 18,122 m^3 of granite, 8 million rivets and the loss of 57 lives, the bridge was completed in 1889. At the opening ceremony on 4 March 1890, the Prince of Wales (later King Edward VII) drove in the last rivet, which was gold-plated and inscribed to record the event.

The Empire State Building is the first skyscraper constructed of steel and was built in 1931. With 102 stories, the building stands a total of 443.2 m tall, including its antenna. The Empire State Building is composed of 60,000 tons of steel with steel columns and beams forming a stable three-dimensional frame throughout the entire structure. The building remained the tallest building in the world for 41 years until 1972 when the World Trade Center claimed this distinction. Today, despite being surpassed in height by many other buildings, the Empire State Building remains an internationally known icon and arguably the most famous building ever constructed.

By now, there are so many large-scale bridges, towers and buildings constructed of steel. Too many to count. Whilst people enjoy the success of steel construction, in particular, the high strength and easy workability of steel, the issues associated with its use in natural environments are looming. It was not until the 1960s that these issues drew appropriate attention from first, perhaps, engineers and then owners and asset managers to now all stakeholders of steel construction. The most daunting issue is arguably corrosion, which is the subject of this book.

1.2 SIGNIFICANCE OF THE BOOK

The significance of the knowledge presented in the book can be demonstrated by many examples of structural collapse initiated by corrosion, the cost associated with corrosion and current practice on corrosion assessment. It is acknowledged that corrosion may not be the only cause for the collapse but at least one of the contributing factors and triggers, and in most cases, the dominant one. It is also acknowledged that mistakes and failures are part of the civilisation of humankind. This is no exception for the development of knowledge on steel corrosion. The point is that when something occurs repeatedly, there is a need to do different things or in a different way. This is the real significance of the book.

1.2.1 Corrosion-induced failures

One of the most publicised structural collapses caused by steel corrosion is perhaps that of the Silver Bridge, which was located in Point Pleasant, Mason County, West Virginia, United States. The bridge was constructed in 1927 and came into service in 1928 (Chen and Duan 2014). The bridge was an eyebar-chain suspension bridge made of steel. The chain was made of heat-treated carbon steel with an ultimate strength of 720 MPa (Lichtenstein 1993). On 15 December 1967, the Silver Bridge suddenly gave in without warning. The eyebar chain ruptured first, and then the entire bridge collapsed. Forty-six people lost their lives, and many more were injured during the collapse. Amongst many reasons that caused the rupture of the eyebar chain, corrosion, in particular, stress corrosion, and fatigue, in particular, corrosion fatigue, should be the ultimate culprits. Naturally, the bridge or more specifically the eyebar chains were exposed to moist air above the Ohio river which are the catalyst for corrosion. The high level of applied stress in the eyebar (with a strength of 720 MPa) resulted in the stress corrosion. Also, the traffic loads on the bridge subjected the eyebars to cyclic loading, which resulted in corrosion fatigue. As would be discussed later in this book, corrosion would also reduce the strength of corroded steel. Both the stress corrosion and corrosion fatigue would cause the cracking of corroded eyebar and accelerate the rupture of the steel bar, leading to ultimate collapse of the bridge (Lichtenstein 1993).

If people think the corrosion-induced collapse of structures is the past, the following example would make them think again. A building at a shopping centre collapsed in 2012. This happened at a shopping centre in Elliot Lake, Ontario, Canada which was built in 1970–1980 with retail space, offices and a parking area. The concrete parking deck, e.g., slabs, was supported by steel beams which sat on steel columns. The beams and columns were connected by bolts. On 23 June 2012, failure occurred at the connection, leading to the collapse of the entire parking deck as shown in Figure 1.2 (adopted from Nastar and Liu 2019). Two people lost their lives, and 22 people were injured. The cause for the collapse was again steel corrosion. The parking deck was constructed of poor-quality concrete which resulted in many cracks after decades of service. These cracks broke the waterproofing system of the deck, and then water leaked to the steel beams and columns which was particularly serious at one of the beam-column connections. As a result, the steel plate at the connection corroded and lost its strength. With the loading from the cars, the connections were exposed to simultaneous stress and corrosion environment, which further accelerated corrosion, leading to the failure of the connection and ultimate collapse of the deck.

Comparing the collapse of the shopping centre car park in 2012 with that of a bridge in the 1960s, it is very surprising or not surprising to find out that, after 50 years, the cause for the ultimate collapse is the same – corrosion. As it is known, the capacity of a structural member is determined by its cross section and material strength. If the corrosion only reduces the cross-sectional area as widely known, the structure would not collapse since this reduction can be determined relatively easily on site through routine maintenance. The fact of this structural collapse suggests that corrosion may reduce the mechanical strength of steel as well, the knowledge of which is less known.

Figure 1.2 Collapse of a parking deck.

Clearly, the knowledge on corrosion and its induced degradation of mechanical properties of corroded steel is imperative to ensure the safe and reliable operation of steel structures during the whole of their service life.

1.2.2 Corrosion-induced costs

It may be true that corrosion-induced collapses are, in the end, very rare events, in particular, with the advanced technology and computing powers for structural design and accumulative knowledge on corrosion. To ensure the safe and serviceable operation of steel structures, maintenance and repairs for corrosion are essential. Corrosion costs a nation's economy substantially in maintenance and repairs of steel structures. According to a report by NACE, the global cost of corrosion is estimated to be US$2.5 trillion, which is equivalent to 3.4% of the global GDP (2013 figure). This is astounding. These costs typically do not include individual casualties or environmental consequences. Through near misses, incidents, forced shutdowns (outages), accidents, etc., several industries have come to realise that lack of corrosion management can be very costly and that, through proper corrosion management, significant cost savings can be achieved over the lifetime of an asset. The knowledge of corrosion and, in particular, its effects on degradation of mechanical properties can help engineers to assess the safety of the corroded steel structures and asset managers to decide when and where to maintain and repair the structure.

Using bridges as an example, 26.6% of bridges in the United States were reported to be structurally deteriorated in 2004 (Li et al. 2018). In Australia, approximately 70% of bridges built before 1985 are subjected to serious corrosion-induced deterioration (Rashidi and Gibson 2012). The total cost for maintenance and rehabilitation of corroded bridges in Australia between 2010 and 2011 was A$1.2 billion, which is an increase of 67% from 2000 (GHD 2015). The cost for maintaining the famous Sydney Harbour Bridge (Figure 1.3) from corrosion damages is about $15 million per year (Haynes 2010).

Last but not least, it should be noted that climate change-induced temperature rise would further exacerbate the severity of corrosion. As it is known, corrosion rate

Figure 1.3 Sydney Harbour Bridge.

would increase with the increase of surrounding temperature. It estimated that 1°C increase in temperature would increase the corrosion rate by up to 0.19 mm/year (Li 2018). This is quite significant. All these suggest that the knowledge on corrosion and its effect on degradation of mechanical properties is imperative to the safety and security of the community and society.

1.2.3 Current practice

The current situation in engineering practice for most design and assessment of steel structures is usually based on the sacrificial approach. In this approach, a commonly acceptable corrosion rate for a given environment, such as thickness reduction of 0.1 mm/year, is adopted to account for the decrease of steel capacity due to corrosion in that environment. With the acceptable corrosion rate, the cross-sectional area of the steel members reduces accordingly in the calculation of capacity and other design parameters of the steel members. In most cases, the corrosion rate is assumed constant, but for some cases, the corrosion rate varies with time. This is achieved by collecting information from the known corrosion damages or loss from a steel structure in a similar environment. The sacrificial approach can be a pragmatic approach for structural design and assessment in corrosion-prone environments. Depending on the information available, it can be conservative, and in some cases, non-conservative, the latter of which would be disastrous as has been presented in Section 1.2.

Intuitively, it is natural to think that, in addition to area reduction of cross section of steel members, there is something else in play in the decrease of steel capacity caused by corrosion. Otherwise, the corrosion-affected structures would not have collapsed due to corrosion because the corrosion-induced reduction of cross-sectional area had been accounted for in the calculation of strength of steel members and in most cases, very conservatively. From the perspective of structural engineers, who design and assess the steel structures, one may wonder if corrosion would affect the mechanical strength of the corroded steel. The purpose of this book is to provide some evidence and base to search for answers and develop knowledge.

A simple test can demonstrate whether corrosion affects the strength of corroded steel or not (see Chapter 3 for details). A small steel plate of 50×10 mm is used as a specimen for tensile strength test. The intact specimen is tested first and then exposed to a corrosive environment. Corrosion is assumed to be uniform and measured by the thickness loss of the plate. The original area of cross section is $50 \times 10 = 500\,\text{mm}^2$. With the thickness loss of 0.5 mm after corrosion, the cross-sectional area is reduced to $50 \times 9 = 450\,\text{mm}^2$, i.e., 10% reduction. The test on intact, i.e., original specimen, gives the ultimate load of 125 kN. If the corrosion only reduces the area of cross section, the second test, i.e., after corrosion, on the specimen would give the ultimate load of 112.5 kN, i.e., the same 10% reduction, but instead, the second test on the corroded specimen gives the ultimate load of 107 kN, i.e., a further reduction of about 5%. If the first reduction is due to the corrosion-induced thickness loss, i.e., area reduction, the second reduction can only be induced by the material itself. By conventional view, corrosion does not affect the mechanical property, i.e., tensile strength, but this simple test does not support the conventional view. With the principle that one test is sufficient to falsify a theory, the corrosion must have affected the tensile strength of steel, which reduces the ultimate load in the second test.

This simple test provides convincing evidence of the need to understand and grasp how corrosion affects the mechanical properties of corroded steel and importantly the need to obtain and establish knowledge on degradation of mechanical properties of corroded steel. This knowledge can then be used to predict and prevent corrosion-induced failures of steel and steel infrastructure, thereby saving money, resources and most significantly the lives of public. It is in this regard that the present book is in order.

1.3 PURPOSES OF THE BOOK

Corrosion of steel has been recognised as the most dominant factor for steel failures, both as a material and in a steel structure. It is a global problem that taxes engineers and researchers alike to come up with solutions. Corrosion of steel reduces its strength and subsequently service life of corroded steel structures, leading to ultimate structural collapse. Almost no steel is immune from corrosion simply because it is exposed to moist air. Given the inevitability of corrosion, it is time to consider a paradigm shift in the body of knowledge of steel corrosion from its diagnosis and protection to prediction and prevention of its damages and consequences, as to be presented in this book.

Prediction of corrosion damages to steel and steel structures has been a daunting and lasting challenge. Considerable research has been conducted, but structures (bridges, transmission towers, pipelines, buildings and railways) essential to a nation's economy, society, environment and wellbeing continue to collapse due to corrosion with catastrophic consequences. The reoccurrence of corrosion-induced collapses of steel structures clearly demonstrates that the existing knowledge on corrosion is inadequate – the corrosion damages to steel have not been correlated to degradation of its mechanical properties. Lack of knowledge in degradation of mechanical properties of corroded steel can lead to overestimation of structural strength and hence overestimation of service life of steel structures. This can be very serious in terms of safety of both structures and communities. This book aims at providing such a knowledge base.

The aim of this book is to present the most recent research on corrosion of steel, including cast iron and ductile iron, and its effect on degradation of its mechanical properties. The emphasis is on corrosion-induced degradation of mechanical properties of steel, such as tensile strength, fatigue and fracture toughness. Knowledge on corrosion-induced degradation of mechanical properties is largely unavailable but much needed to accurately assess and predict failures of steel structures due to corrosion. This knowledge is also imperative in planning for maintenance and repairs of corrosion-affected steel structures.

The knowledge to be presented in the book covers the basics of corrosion science of steel. Models for corrosion-induced degradation of mechanical properties are also presented in the book with a view of wider practical applications. These models can provide guidance to engineers and asset managers in accurately assessing and predicting corrosion-induced failures of steel and steel structures. As to be illustrated in Section 1.2, the cost related to corrosion-induced maintenance, repairs and unexpected collapses is huge, in particular, when the collapses lead to the causalities which are beyond estimation. Therefore, the knowledge that can be used to prevent corrosion-induced failures could bring about immense benefits to the industry, business, society and community.

This book will focus on corrosion effect on the degradation of mechanical properties which is examined at both macro level, e.g., the mechanical strength, and micro level, e.g., element composition, and also with stress and without stress. Given that corrosion has been a "well-trodden" topic for many decades, this book takes a quite different approach to deal with it. The approach to be taken in the book is more mechanistic, focusing on physical process and damages, more phenomenological, focusing on observations, and more practical, focusing on applications of the developed knowledge, and yet without losing insightful analysis of fundamental causes of degradation of mechanical properties of corroded steel. As a "golden rule" of material science, the structure dictates the property of the material. The microstructure of the corroded steel will also be analysed with a view to explore fundamental causes for degradation of mechanical properties of corroded steel. It is believed that this approach has the appeal to a much wider pool of users from material scientists to structural engineers, from academic researchers to practical designers and from university lecturers to students.

In the preface of their book (Revie and Uhlig 2008), the authors state as follows:

> Although the teaching of corrosion should not be regarded as a dismal failure, it has certainly not been a stellar success providing all engineers and technologists a basic minimum "literacy level" in corrosion that would be sufficient to ensure reliability and prevent failures.

Clearly, the corrosion-induced degradation of mechanical properties of steel cannot be more of "literacy level" in ensuring reliability and preventing failures of steel and steel structures. This reinforces the need of the book presented here.

The outline of the book is as follows. Following the Introduction, the basics of steel corrosion will be presented. In Chapter 2, emphasis will be on those aspects that are closely related to the mechanical properties of steel, such as the composition of chemical elements, the factors that affect the corrosion and mechanical properties, and the test methods on corrosion

Chapter 3 will discuss the effects of corrosion on the mechanical properties of steel in a phenomenological manner. Tests on corrosion in laboratories and on sites will be presented. The focus of the chapter will be on degradation of the tensile properties of corroded steel from both micro- and macro-structural points of view. The degradation of other two important mechanical properties, namely fatigue strength and fracture toughness, of the corroded steel will also be covered.

Corrosion effects on the mechanical properties of cast iron and ductile iron will be presented in Chapter 4. Simulation of corrosion in both real soil and soil solution will be discussed. Since the cast iron is a very brittle metal, the chapter will focus on the degradation of tensile strength and fracture toughness of corroded cast iron with analysis at both micro- and macro-structural levels. The degradation of tensile strength and fracture toughness of corroded ductile iron will also be included.

Chapter 5 will discuss other corrosion-induced damages, including delamination and hydrogen embrittlement. It will first discuss the stress effect on corrosion with comparison of the difference in corrosion with and without stress. Emphasis will be on preferred corrosion, its mechanisms and its effect. Control and prevention of preferred control will also be presented. Hydrogen embrittlement and its effect and mechanism will also be discussed.

Chapter 6 will cover applications of corrosion knowledge presented in this book to practical assessment of corroded structures, using practical examples. The concept of developing an acceleration factor as a tool of calibration is presented with a view to facilitate the application of test results presented in the book. To conclude the book, future research and development in steel corrosion will be presented with focus on a very novel idea to develop a new theory of corrosion-induced degradation of mechanical strength of corroded steel, tentatively called *Nanomechanics of Steel Corrosion*.

Chapter 2

Basics of steel corrosion

2.1 INTRODUCTION

Steel is perhaps the most widely used building material in terms of its strength and versatility in many aspects. However, the advantages of steel as a building material over other counterparts are hammered by its tendency to revert to its natural state, i.e., from where it is extracted. This is one of the root causes of steel corrosion. Some conditions only serve as a catalyst for corrosion to initiate and progress slower or faster. To grasp corrosion and its effect on steel and the mechanical properties of steel, it is better to start with steel itself: how steel is manufactured, what chemical composition it consists of and what mechanical properties it possesses. This knowledge can help understand corrosion of steel more profoundly and appreciate its effects on steel, in particular mechanical properties of steel, more accurately. Since this book focusses more on the effect of corrosion on the degradation of mechanical properties of corroded steel from a mechanistic perspective, more details of corrosion science and related chemical reactions can be referred to in other books that are more focussed on them, such as Revie and Uhlig (2008) and Marcus (2011).

2.1.1 Making of steel

Steel is an alloy of iron with other chemical elements and consists mainly of iron (Fe) as the base material and other minor elements of which carbon (C) takes a large proportion. Steel is made in an integrated plant by one of two processes: blast furnace and electric arc furnace. Blast furnaces use mainly raw materials (iron ore, limestone and coke) with some scrap steel to make steel, whereas electric arc furnaces use mainly scrap steel. The most common process for steel-making is the integrated steel-making process via blast furnace – the basic oxygen furnace.

Essentially, there are three stages in making steel via blast furnace. In stage one, iron ore, mainly hematite (Fe_2O_3) and magnetite (Fe_3O_4), is mixed with coke, a solid porous fuel with a high carbon content, in the furnace as schematically shown in Figure 2.1 (www.thermofisher.com). The mixture is heated to about 1,000°C, known as sinter, and then limestone ($CaCO_3$) is added in. This is the raw material for making steel. At this point, there are many impurities in the raw material, which have to be removed to ensure the steel is not brittle. In stage two, hot air, up to 1,900°C, is blown into the furnace through nozzles in the lower section. The hot air burns the coke,

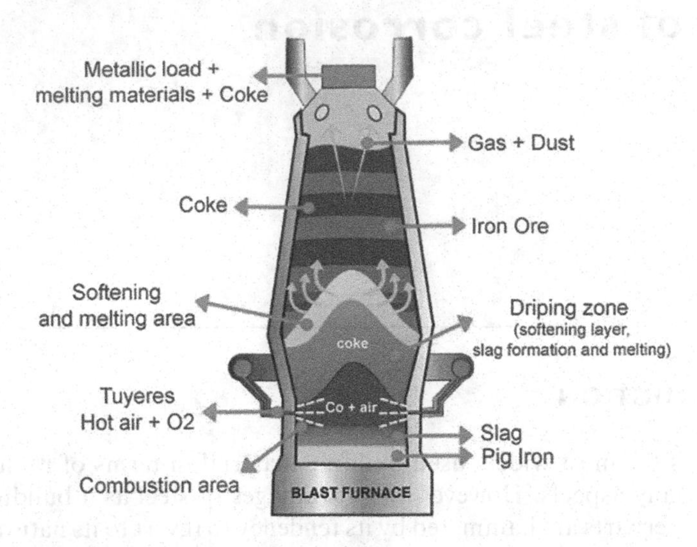

Figure 2.1 Steel making in blast furnace.

which produces carbon monoxide (CO). The hot air also melts the iron ore, and carbon monoxide reacts with the iron ore, together becoming molten iron which sinks to the bottom of the furnace. In the meantime, limestone reacts with impurities in the iron ore which becomes slag, floating on top of molten iron. After plenty of such reactions, molten iron and slag flow out through the tap holes at the bottom of the furnace. The final stage is to remove or control the impurities in the molten iron, which are mainly carbon (C), manganese (Mn), silicon (Si), phosphorus (P), sulphur (S) and a very small quantity of other elements. This is the raw product of steel or primary steel making. At this stage, other chemical elements, such as nickel, chromium, molybdenum and titanium, are added in the molten iron or hot metal to make different types of steel with different mechanical properties.

The final products of steel that are used in engineering, such as construction or other fields, are manufactured by two main casting processes. One process is continuous casting, which is to pour molten steel into cooling moulds to form various products as required. This allows the steel to become hard, and the steel is drawn out whilst it is still hot, which is the so-called hot-rolled steel. The other process is ingot casting, which produces semi-finished steel. Ingots can be heated and formed by repeated hammering, such as blacksmith, to forge various shapes of steel products, such as steel rings. Another process of making steel products is cold formed. Cold-formed steel is made from strips of quality sheet steel that are fed through roll forming machines with a series of dies that progressively shape the steel into a variety of shapes, including "C", "U" and "Z" sections.

2.1.2 Chemical composition of steel

As an alloy, iron is the base material of steel, and other elements are in small quantities. The standard chemical composition of steel is iron (Fe, 90%–95%), carbon (C,

<4.5%) and other elements, mainly manganese (Mn), silicon (Si), phosphorus (P) and sulphur (S). The presence of 4.5% carbon makes steel brittle. Addition of other elements will make different types of steel. In general, steel is categorised into four groups by composition (www.meadmetals.com): carbon steel, alloy steel, stainless steel and tool steel. Carbon steels only contain trace amounts of other elements besides carbon and iron. This group is the most commonly used steel, accounting for 90% of steel production. Carbon steel is further divided into three subgroups, depending on the amount of carbon in the metal by weight. Low-carbon steels contain up to 0.2% of carbon content. Wrought iron contains the least carbon, usually less than 0.1%. Low-carbon steel is also known as mild steel, which is mostly for structural use with a carbon content of 0.15%–0.2%. Medium-carbon steels contain a carbon content between 0.3% and 0.6%, which are not much used in construction. High-carbon steels usually contain more than 0.6% carbon in the metal. Eutectoid steel is high-carbon steel with carbon content greater than 0.8%. Another type of high-carbon steel is cast iron and ductile iron, which contain more than 1.67% carbon. This book will focus on structural steel, i.e., mild steel with a carbon content between 0.15% and 0.2%. It also covers cast iron and ductile iron; together they are called ferrous metals or in short, steel.

Other types of steel include alloy steels, which contain alloying elements like nickel (Ni), copper (Cu), chromium (Cr), tin (Sn), aluminium (Al) and so on. These additional elements are used to alter the mechanical properties of steel, such as strength and ductility, and other properties such as corrosion resistance and machinability. They are widely used in heavy-duty structures and machineries. Stainless steels contain 10%–20% chromium (Cr) as their alloying element and are valued for their high corrosion resistance but at a higher cost as well. These steels are commonly used in medical equipment, piping, cutting tools and food processing equipment. Tool steels make excellent cutting and drilling equipment as they contain tungsten (W), molybdenum (Mo), cobalt (Co) and vanadium (V) to increase heat resistance and durability and prevent tear and wear commonly experienced in using tools. Test certificate is required for any steel products before they come into use. To do this, samples are taken from the liquid steel to check the composition of the required elements in % by weight. In addition to iron (Fe), five main basic elements that need to be specified are carbon (C), manganese (Mn), silicon (Si), phosphorus (P) and sulphur (S). Each element is specified for a specific range for different requirements of mechanical properties of steel.

As a summary, the influence of chemical elements on the mechanical properties of steel is shown in Table 2.1. It can be seen that there is a strong correlation between the chemical composition and mechanical properties of steel. This means that if the element contents change, the mechanical properties would change accordingly or at least be affected. For example, a reduction of iron content of steel can degrade its tensile strength, fatigue resistance and fracture toughness. This can be understood since iron is the main element of steel. If the corrosion reduces the iron content, e.g., due to consumption during electrochemical reactions, the tensile strength, fatigue resistance and fracture toughness of steel may be reduced accordingly. Likewise, if the sulphur content reduces, e.g., due to reaction with other elements, all mechanical properties of steel can increase. It can be also noted from the table that the effect of some elements is not the same on all mechanical properties; some are positive and some negative. Of all, the most noticeable element is carbon, which affects the tensile strength positively but negatively for fatigue resistance and fracture toughness. This appears to be well

Table 2.1 Effect of Chemical Elements on Mechanical Properties of Steel

Element	*Tensile Strength*	*Fatigue*	*Fracture Toughness*
Iron (Fe)	Base element with a positive effect on all mechanical properties		
Carbon (C)	Positive	Negative	Negative
Manganese (Mn)	Positive	Positive	Positive
Silicon (Si)	Positive	Positive	Negative (maybe)
Phosphorus (P)	Positive when <0.1%	Negative (for both < and > 0.1%)	Negative when >0.1%
Sulphur (S)	Negative	Very negative	Very negative
Chromium (Cr)	Positive	Positive	Negative
Nickel (Ni)	Positive	Positive	Positive

understood. In addition, it needs to be noted that there can be interactions amongst elements, which may change its effects on mechanical properties. Exactly how these interactions affect the mechanical properties of steel needs to be established in another book.

2.1.3 Mechanical properties of steel

To understand the corrosion effect on the mechanical properties of steel, it is first necessary to know the engineering properties used by practitioners. Of many such properties, this book focusses only on three important properties that have to be used in design and assessment of steel structures and in the meantime are affected by corrosion, which is the theme of the book. These properties are tensile strength, fatigue and fracture toughness. The tensile property of steel is determined from the tensile tests, which are the basic material tests for steel and indeed other engineering materials. The test produces a stress-strain curve, denoted by $\sigma - \varepsilon$ curve, from which three basic mechanical properties can be determined:

- Yield strength: which indicates the limit before which the steel behaves elastically.
- Ultimate strength: which indicates the maximum load that the steel can carry.
- Failure strain: at which steel ruptures or snaps.

These three properties will be used for studying the corrosion effect in later chapters.

Fatigue is, in general, the deterioration of a material's resistance to the progressive and localised damage when the material is subjected to cyclic loading. The fatigue resistance of steel is represented by the *S-N* curve, which is a plot of the magnitude of stress range (S) versus the number of load cycles (N) to failure. This can be expressed as $N = AS^{-B}$, where A is the fatigue strength coefficient, and B is the fatigue strength exponent. Values of A and B are specified in standards for specific fatigue classifications, based primarily on experiments. In practice, the *S-N* relation is usually expressed in log scale as $\log N = \log A - B \log S$ (Zhao et al. 1994). For any specific mild steel subjected to normal stress range, the *S-N* curve can be determined by knowing the fatigue classification of the steel. Steel experiences fatigue under cyclic, normal and shear stresses. Unlike the tensile strength of steel, the *S-N* curve for shear fatigue cannot be simply related to that for tensile (normal stress) fatigue even though it is also governed by the same relation, i.e., $N = AS^{-B}$. Once the parameters A and B are determined, an S-N curve can be easily established for shear fatigue. More information of fatigue is presented in Section 3.4.1.

Fracture toughness represents the ability of steel to resist fracture. It has become an important mechanical property of steel in the engineering design and integrity assessment of steel structures, in particular, corrosion-affected steel structures where defects, such as corrosion pits, can initiate cracking. Fracture toughness can only be determined from fracture toughness tests prescribed in standards. Whilst the tensile properties of steel can be relatively easy to obtain, the fracture toughness of steel is difficult to determine even with the advanced testing facilities. To make the matter worse, different standards used can produce different values for the fracture toughness of the same steel (Gao et al. 2020).

2.2 CORROSION PROCESS OF STEEL

Corrosion is a natural chemical process not exclusive to only steel or even metals. Non-metallic materials can also corrode. In simple terms, corrosion is the oxidation of a material with air, usually moist air, or reaction of this material with oxygen usually in the presence of water. In general, corrosion of steel is a very complex process affected by many factors, mainly the surrounding environment and the steel material itself. The complexation is however much more in details of the chemical process rather than in principle. Since the most commonly exposed environment for steel structures is mainly atmosphere, and that for cast iron is soil (such as underground pipes), this book focusses more on the atmospheric corrosion for steel and corrosion in soil for cast iron and ductile iron. Furthermore, this book is intended for engineers for assessing corrosion-induced damages to steel and steel structures; fundamentals of corrosion science and chemical reactions will not be presented in detail here since they can be easily found in other books, such as Revie and Uhlig (2008) and Marcus (2011).

2.2.1 Electrochemical reactions

In essence, steel corrosion is an electrochemical process that occurs when two or more points on the steel surface have a potential difference, and two chemical reactions, i.e., oxidation and reduction, take place simultaneously on the steel surface (Cramer and Covino 2003). The process of steel corrosion involves four basic parts:

- A metal, which acts as an electrical conductor for anodic and cathodic electronic transfer;
- An anode, where electrochemical oxidation takes place and electrons are liberated;
- A cathode, where electrochemical reduction occurs, and electrons transferred from the anode are consumed;
- A conductive medium, which is the aqueous medium or electrolyte or the local environment the steel is exposed to.

The oxidation reaction is also known as anodic reaction, in which electrons are lost, resulting in a non-metallic state. Anode is the point where electrons on the steel surface are lost and corrosion takes place. The oxidation reaction can be expressed as follows (Revie and Uhlig 2008):

$$Fe \rightarrow Fe^{2+} + 2e^{-} \tag{2.1}$$

The reduction reaction is known as cathodic reaction, which balances the anodic reaction. In this reaction, ions in the electrolyte accept electrons that are released from the electrically connected anode. There are two types of cathodic reactions – hydrogen evolution and oxygen reduction (Chalaftris 2003). The availability of oxygen in the environment determines which reaction plays a dominant role. In an environment where there is limited oxygen (i.e., in deaerated solution), the cathodic reaction can be expressed as follows:

$$2H^{+} + 2e^{-} \rightarrow H_2 \tag{2.2}$$

This reaction proceeds rapidly in an acidic environment. In alkaline environments and atmosphere, the dissolved oxygen accelerates the catholic reaction by causing oxygen reduction reaction, which can be expressed as follows (Revie and Uhlig 2008):

$$O_2 + 4H^{+} + 4e^{-} \rightarrow 2H_2O \text{ (in acidic solution)} \tag{2.3a}$$

$$O_2 + 2H_2O + 4e^{-} \rightarrow 4O \text{ (in neutral/alkaline solution)} \tag{2.3b}$$

From Equations (2.1)–(2.3), it can be seen that at anode, electrons are produced which are transported via electrolyte or aqueous medium to the cathode where they are consumed with oxygen and water. Ferrous and hydroxyl ions ($Fe^{2+}$and OH^{-}) flow within the aqueous medium so that they can react with each other. The progress of corrosion, or corrosion rate, is controlled by the cathodic reaction.

With more dissolved oxygen and water in the aqueous medium, further reactions can lead to the formation of corrosion products. Adding Equations (2.1) and (2.3) together, this reaction can be expressed as follows:

$$2Fe+2H_2O+O_2 \rightarrow 2Fe(OH)_2 \tag{2.4}$$

where the product $Fe(OH)_2$, i.e., ferrous hydroxide, acts as a diffusion barrier that exists on the steel surface. Diffusion barrier is also known as passive oxide film which is very thin (in nanometres). The oxide film temporarily protects steel from further corrosion.

The ferrous hydroxide (or iron (II) hydroxide), $Fe(OH)_2$, is chemically not stable and susceptible to oxidation, in particular, at the outer surface of the passive oxide film where dissolved oxygen is more accessible. With dissolved oxygen, it converts to hydrous ferric oxide or ferric hydroxide (iron (III) hydroxide), $Fe(OH)_3$, shown as follows:

$$2Fe(OH)_2+H_2O+O_2 \rightarrow 2Fe(OH)_3 \tag{2.5}$$

$Fe(OH)_3$ is also unstable and will break down into hydrated ferric oxides and water soon after its formation, which can be expressed as follows:

$$2Fe(OH)_3 \rightarrow Fe_2O_3 \bullet H_2O + 2H_2O \quad (2.6)$$

The products of steel corrosion is what is known as rust, which is mostly composed of the ferric hydroxide (iron (III) hydroxide), $Fe(OH)_3$ (Equation 2.5) and hydrated ferric oxide (iron(III) oxides), $Fe_2O_3 \cdot H_2O$ (Equation 2.6).

By now, it can be seen that steel corrosion is such a process that the steel tends to return to its original state, i.e., iron ore or hematite (Fe_2O_3), from which it is extracted. This tendency is natural and can be the root cause of corrosion for steel. The surrounding environment, such as the aqueous medium, only acts as a catalyst that initiates and accelerates the corrosion process. From a metallurgic point of view, metals are in high free energy state when they are extracted. There is a tendency for them to corrode and return to their original state because this is the natural and stable state for them.

The corrosion process (Equations 2.1–2.6) repeats itself between anode and cathode on the steel surface, which, subsequently, causes the dissolution of the steel material. Once the mass of the steel is consumed to a certain extent by corrosion, the dimensions of the steel reduce considerably. As such, steel can no longer carry the load it is designed to carry. This means that steel can no longer function as a structural material, resulting in the collapse of the steel structures or end of their service life. This is the conventional view and straightforward. As a summary, the corrosion process is schematically depicted in Figure 2.2.

2.2.2 Progress of corrosion

From the electrochemical reactions of corrosion, i.e., Equations (2.1)–(2.6), it is clear that oxygen and water are two essential elements for corrosion to initiate and progress. The speed of corrosion of any metal depends on environmental conditions as well as the type and condition of the metal. It also depends on the corrosivity of the aqueous medium or electrolyte and variations in the potential of the metal surface due to the creation of anodic and cathodic sites. The corrosion rate of steel, which is defined as the corrosion growth per unit time, is controlled by the cathodic reaction. There are

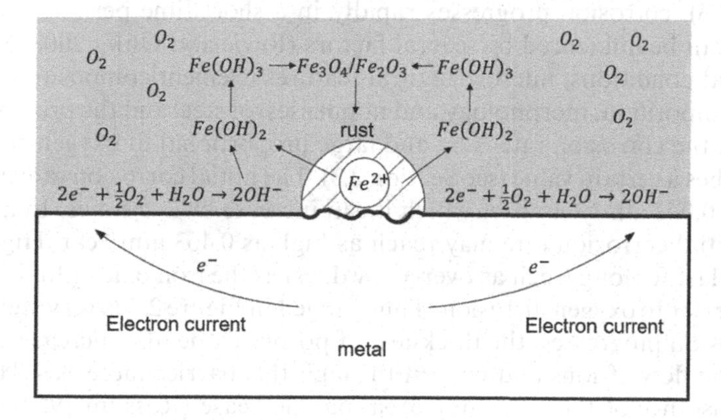

Figure 2.2 Process of steel corrosion and its products.

Table 2.2 Unit of Corrosion Rate

Symbol	*Meaning*	*Comments*
mA/cm^2	Corrosion current in mA per square centimetre	Used more in research
gmd	Corrosion mass loss in gram per square meter per day	Normalised by surface area
mm/year	Corrosion penetration (or thickness loss) in mm per year	Used more in practice
Mpy	Corrosion penetration in milli-inches per year	Used mainly in the USA
Mdd	Corrosion mass loss in milligram per square decimetre per day	Not most used
%	Corrosion loss in percentage	Used more in practice

a few measures of corrosion rate, namely by corrosion current, by mass loss and by penetration, i.e., the dimension loss. Commonly used units for corrosion rate are summarised in Table 2.2.

In general, the corrosion process can be divided in two phases: initiation and propagation. However, for steel without any protection, such as coating or reinforcing steel in concrete, the initiation phase is much shorter compared with propagation and even much shorter compared with the whole of service life of steel structures. As per Equation (2.4), steel is protected from corrosion at the beginning of its exposure in air or other environments. But this protection is short mainly because the passive oxide film that protects the steel from corrosion, i.e., ferrous hydroxide ($Fe(OH)_2$), is chemically unstable and very susceptible to oxidation. Since the initiation time is short, not much research has been undertaken to determine exactly how long it takes for corrosion to initiate in various environments. In the atmospheric environment, the estimated initiation time of steel corrosion is about 6 months to 2 years. In marine environment, the corrosion can initiate in short time if the steel is not protected by any means.

Once initiated, the corrosion progresses relatively fast, depending on the availability of required essential elements for corrosion, i.e., oxygen and water. The speed of corrosion progress is measured or indicated by corrosion rate as shown in Table 2.2 and commonly in mm/year. Conceptually, the progression of steel corrosion can be divided into three stages as schematically shown in Figure 2.3. In the first stage (indicated as I in Figure 2.3), corrosion progresses rapidly in a short time period. The corrosion rate of steel can be influenced by several factors (Revie and Uhlig 2008, Saha 2012) – environmental conditions, microstructural features (element composition, grain size, iron phase composition, morphology and impurities) of steel and the presence of stress. At this stage, the corrosion rate is by and large proportional to oxygen concentration before it reaches a certain value (see Section 2.3). The initial corrosion rate in the atmosphere can be 0.047 mm/year after which it can increase by 2–5 times. In air-saturated water, the initial corrosion rate may reach as high as 0.463 mm/year. High corrosion rate may not last for long (such as over a few days) as the iron oxide film is formed and acts like a barrier to oxygen diffusion. Thus, stage I in Figure 2.3 is very short in reality.

As corrosion progresses, the thickness of porous oxide film increases, and the resistance to the flow of ions and oxygen through this barrier increases. However, the electrical resistance of the corroding area may decrease (Rossum 1969). Also, when this barrier mainly consists of iron oxides, it can act as a site for oxygen reduction

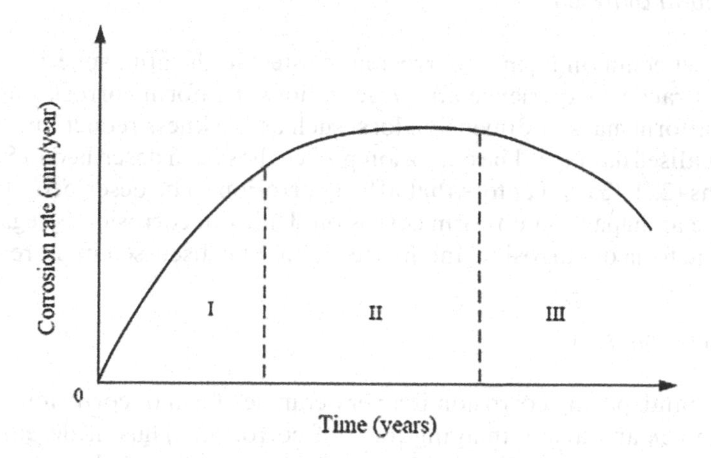

Figure 2.3 Conceptual model of the corrosion process (schematic).

(Stratmann and Müller 1994). The corrosion rate may remain relatively constant. This is the second stage (indicated as II in Figure 2.3) which is a steady-state stage with a corrosion rate of about 0.0463–0.116 mm/year. Stage II can last for a few months or years depending on the surrounding environment. Corrosion rate in this stage is largely influenced by the categories of corrosion products and sulphates, pH and carbonate contents of the exposure environment (Norin and Vinka 2003).

When the composition and physicochemical properties of oxide film change, the corrosion rate may reduce. For instance, when the iron oxides are replaced by iron carbonates, the electrical resistance of this barrier will increase. Considering that the thickness of protective film tends to be thicker and corrosion reaction is diffusion controlled, the corrosion rate is further reduced. This is the third stage in which the corrosion rate is often less than 0.1 mm/year (Marshall 2001). The transition time amongst different stages depends on the environmental conditions (Cole and Marney 2012, Aung and Tan 2004).

2.2.3 Types of corrosion

There are many types of corrosion which are categorised in different ways. Based on corrosion progression, there are two basic forms of corrosion: uniform corrosion and pitting corrosion. Based on the exposure environment, there are atmospheric corrosion, marine corrosion, soil corrosion and so on. Based on the mechanism, there are galvanic corrosion, microbial corrosion, stress corrosion and so on. Galvanic corrosion is an electrochemical process in which one metal corrodes preferentially when it is in electrical contact with another exposed to the same electrolyte environment. It should be noted that although there are many types of corrosion, the fundamental process of corrosion is the same for all types of corrosion once it starts or initiates. This is the electrochemical reactions as described in Equations (2.1)–(2.6). It may be appreciated that it is not possible to describe all these types of the corrosion in this chapter nor is this the purpose of the book. Within the scope of this book, the following types of corrosion are presented.

2.2.3.1 Uniform corrosion

This is the most common form of corrosion for steel in the atmospheric environment based on the practical experience and observations. Uniform corrosion is characterised by the uniform mass or dimension loss, such as thickness reduction, of steel with little or no localised damage. The corrosion process has been described in Section 2.2.1 (see Equations (2.1)–(2.6)). Factors that affect corrosion to be described in Section 2.3 would all have an impact on uniform corrosion. Uniform corrosion is regarded as the most simplistic form of corrosion and hence will not be discussed in more detail here.

2.2.3.2 Pitting corrosion

On the other hand, pitting corrosion is a very complex form of corrosion and perhaps the most complex and also damaging form of corrosion. Thus, it deserves more attention. Pitting corrosion is a localised form of corrosion by which cavities or pits are produced on the surface of steel (Revie and Uhlig 2008). The electrochemical reactions of pitting corrosion are the same as those for uniform corrosion. Pitting corrosion mainly occurs in three steps – the breakdown of passive oxide film, i.e., initiation of pitting corrosion, growth of corrosion pits, and reforming of the passive oxide film (re-passivation). These three steps repeat themselves and lead to the growth of pits as schematically shown in Figure 2.4 (Revie and Uhlig 2008).

Corrosion pits are initiated due to the localised breakdown of the passive oxide film, due to the following mechanisms (Li 2018):

1. Physical damage of the passive oxide film caused by, e.g., scratches on the surface;
2. Ingress of aggressive ions (such as chloride Cl^- and sulphate SO_4^{2-}) at some locations, which creates a low-pH environment and weakens the stability of the passive oxide film;
3. Localised stresses rupturing the passive oxide films;

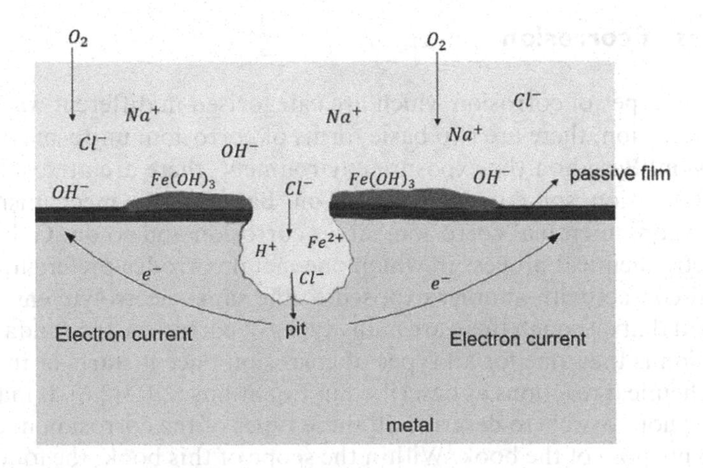

Figure 2.4 Pitting corrosion.

4. Non-homogeneous environment accelerating the dissolution of passive oxide film at some locations;
5. Defects on the surface of steel.

The growth of corrosion pit depends on two potentials at the pit. One is pitting potential which refers to the least positive current and voltage at which pits develop or grow on a metallic surface. The other is re-passivating potential which refers to the critical current and voltage below which pitting corrosion does not occur (Soltis 2015). When the pitting potential exceeds the re-passivating potential, corrosion pit propagates. There are five main factors that can affect the growth rate of pits (Revie and Uhlig 2008), summarised as follows:

1. Chloride concentration: The growth rate of pits increases with the increase of chloride concentration in the aqueous medium since chloride prohibits the reformation of the passive oxide films.
2. Chemical composition of steel: Some alloying elements affect the growth rate of pits. For example, high content of chromium in steel reduces the growth rate of the corrosion pits because chromium contributes to the reformation of the passive oxide film.
3. Microstructure of steel: Corrosion pits can grow faster at the locations where there is a large quantity of impurities, a high proportion of pearlite phases of iron and high dislocation density at grain boundaries.
4. Temperature: Pitting corrosion can only be initiated when the temperature is above a critical pitting temperature (ranging from 10°C to 100°C for steel). Below this temperature, the initiation of corrosion pits requires an extremely high electrochemical potential. Above this temperature, the growth rate of corrosion pits increases with the increase of temperature.
5. Bacteria and pollutants: The presence of bacteria and pollutants at some locations can contribute to localised corrosion and increase the growth rate of pits.

Pitting corrosion mainly initiates due to the rupture of the passive oxide film. Therefore, it is reasonable to assume that pitting corrosion can be more serious under combined stress and corrosion environment (Li 2018). Pits can further initiate or propagate to cracks under combined stress and corrosion environment which significantly increases the risk of fracture of the corroded steel (see Chapter 5).

2.2.3.3 Crevice corrosion

This is similar to pitting corrosion. Crevice corrosion is a form of localised corrosion that occurs inside a metal–metal crevice or metal–non-metal crevice. As the diffusion is restricted to inside the crevice, the oxygen content and pH value are much lower than those outside the crevice. An anode can be created at the region on the metal surface, as the oxygen content is low in the crevice. As a result, positively charged ions appear in the crevice, and the stagnant solution becomes acidic. Also, if chlorides are present in the crevice, a local and small galvanic cell would be created, which accelerates the crevice corrosion.

2.2.3.4 *Microbial corrosion*

This is also called bacterial corrosion or microbially induced corrosion, which occurs with the involvement of microorganisms or bacteria. These bacteria can be classified as aerobic (requires oxygen) and anaerobic (requires no or little oxygen). Microbial corrosion affects almost all types of alloys, such as carbon steels, stainless steels, aluminium alloys and copper alloys. The biological activities of these bacteria tend to create a biofilm and colonise it, which produces a radically different environment, such as pH, the concentration of ion and oxygen as compared to the surrounding global environment. As a result, the electrochemical process is modified, often with an accelerated rate of corrosion. For example, bacteria like *Acidithiobacillus* can produce sulphuric acid and bacteria *Ferrobacillus ferrooxidans* can oxidise iron directly into iron oxides and iron hydroxides (Chen 2018). Bacteria corrosion can also appear in the form of pitting corrosion.

2.3 FACTORS AFFECTING CORROSION

There are a considerable number of factors that affect corrosion, which can be classified in many ways. In general, there are two categories of factors which are internal and external. Internal factors mainly consist of material composition, impurities and defects. Different proportions of various chemical elements in steel, such as carbon, phosphorus, sulphur and silicon, have different impacts on corrosion. External factors mainly include environmental factors and physical factors. Environment mainly refers to aqueous medium, or electrolyte, through which electrochemical reactions take place. Such environmental factors include dissolved oxygen, temperature, pH value of electrolyte solution, dissolved salts (such as chloride and sulphate) in the solution as well as the presence of microorganisms (such as microalgae, bacteria and fungi). Physical factors mainly include stress, fatigue and pressure that cannot be ignored as factors affecting corrosion which will be discussed in other chapters.

As stated in Chapter 1, two important types of corrosion and its effect on degradation of mechanical properties of corroded steel will be discussed in this book. One is atmospheric corrosion which implicates almost all steel structures built above ground. The other is the corrosion in soil which is mainly for underground structures in particular pipes. Therefore, this section focusses on how environmental factors, material factors and soil factors affect corrosion of steel in a more qualitative manner than quantitative. More details on factors affecting corrosion can be referred to in other published literature such as Revie and Uhlig (2008).

2.3.1 Environmental factors

2.3.1.1 *Concentration of dissolved oxygen*

Oxygen, mostly in its dissolved form in the aqueous medium, or electrolyte solution, is the essential element in corrosion reactions as clearly shown in the electrochemical reactions, (see Equations (2.3)–(2.6)). In the absence of dissolved oxygen, the corrosion rate at room temperature is negligible. The amount of dissolved oxygen at the corrosion site affects corrosion positively. Initially, the increase in oxygen concentration increases the corrosion rate of steel by accelerating the cathodic reaction as indicated by Equation (2.3a). When the concentration of dissolved oxygen is below about 12 mL/L, an increase of oxygen concentration of 2 mL/L will lead to an increase of corrosion rate of 0.062 mm/year (Revie and Uhlig 2008). Once the concentration goes beyond 12 mL/L, the corrosion rate starts to decrease rapidly and then plateau to a low value. The decrease of corrosion rate is due to re-passivation of the steel surface by oxygen as indicated by Equation (2.4). The corrosion rate decreases with the increase of oxygen concentration since there is more oxygen at the steel surface than that can be consumed by corrosion reactions. The excess oxygen then forms a passive film which prevents the steel from further corrosion.

2.3.1.2 *Temperature*

Temperature affects steel corrosion positively both in aqueous environment and atmospheric environment. It was reported (Revie and Uhlig 2008) that corrosion rate at a given oxygen concentration approximately doubles for every 30°C increase in temperature when corrosion is controlled by diffusion of oxygen. When the corrosion process involves hydrogen evolution, the increase of corrosion rate can be more than double for every 30°C rise in temperature. The rate for iron corroding in hydrochloric acid, for example, approximately doubles for every 10°C rise in temperature. In general, corrosion rate increases when the temperature rises to about 80°C and then decreases with further increase of temperature. The increase in corrosion rate with the rising temperature below 80°C can be explained with the law of chemical kinetics. When the temperature increases, the diffusion speed of oxygen to steel surface is accelerated, and the diffusion speed of ferrous oxide (FeO) and ferrous hydroxide ($Fe(OH)_2$) in the electrolyte solution is also accelerated. As such, the resistance of the electrolyte is reduced, which accelerates electronic current and hence the corrosion rate (Hou 2016; Revie and Uhlig 2008). The decrease of corrosion rate above 80°C may be related to the significant fall of oxygen solubility in water with the rising temperature. This effect eventually exceeds the accelerating effect of temperature. As a result, the excess oxygen forms the passive film that prevents steel from rapid corrosion.

2.3.1.3 *Relative humidity*

Relative humidity (RH) is a critical factor for atmospheric corrosion. In this case, a thin aqueous layer is formed on the metal surface (Saha 2012, Revie and Uhlig 2008), which depends on the deposition rate of the air pollutants and varies with the wetting conditions. Once the RH reaches a critical value, the thin aqueous layer forming on the oxidised metal leads can be considered as electrolyte containing dissolved oxygen which triggers a range of redox reactions. In general, RH affects the corrosion positively, especially when the ambient RH reaches a critical level, about 60% for steel (Revie and Uhlig 2008). When RH is under 60%, the effect of humidity on corrosion is negligible since the atmospheric corrosion occurs in the presence of the thin aqueous layer that forms on the oxidised steel after the RH reaching the critical value (Evans 1960). The reason that the RH has little effect on corrosion rate before it reaches a critical value can be attributed to the solid oxide film formed spontaneously on the metal surface (Saha 2012; Revie and Uhlig 2008). Ferrous hydroxide $\left(Fe(OH)_2\right)$ and hydrated ferrous oxide ($FeO.nH_2O$) are the first diffusion barrier layer formed on the surface. The oxide film can reach a maximum thickness of 1–5 nm, preventing further corrosion reactions (Brockenbrough and Frederick 2011). However, when RH reaches a critical level, a thin electrolyte moisture layer forms on the steel surface. As such, the corrosion rate increases because the oxide film is no longer protective as air and water can still penetrate the rust via the electrolyte layer (Saha 2012).

2.3.1.4 pH value

The pH value of electrolyte solution has a positive effect on corrosion rate. In general, a reduction of pH, i.e., increase of acidity, leads to the acceleration of corrosion rate (Revie and Uhlig 2008). However, when pH is greater than 5, the corrosion rate is almost independent of pH and instead depends more on other factors, such as oxygen concentration and its diffusion, temperature and so on. When the pH value is less than 4, the passive protective film on the surface of steel can be dissolved. This allows iron to come into almost direct contact with the electrolyte solution (Revie and Uhlig 2008, Saha 2012) and hence increases the corrosion rate. The presence of some microorganisms in the exposure environment (such as sulphate-reducing bacteria) can change the pH which subsequently accelerates corrosion rate (Petersen and Melchers 2012).

2.3.1.5 Salts

Various salts exist in the electrolyte solution, in particular sodium chloride (NaCl) and magnesium chloride ($MgCl_2$). Generally, chloride affects the corrosion positively before its concentration reaches a threshold, about 3% (by weight of solution) (Revie and Uhlig 2008). This is why steel exposed to a marine environment usually has a higher corrosion rate than that in a rural environment due to chloride ion erosion. After that threshold, corrosion rate decreases. This is because the oxygen solubility in the solution decreases continuously with the increase of chloride concentration. Therefore, lower chloride concentration will result in higher corrosion rate. It is generally considered that sodium hydroxide (NaOH) formed by oxidation reaction at the cathode stie does not react immediately with iron dichloride $\left(FeCl_2\right)$ formed at anodes. Instead, these substances diffuse into the solution and react to form ferrous hydroxide $\left(Fe(OH)_2\right)$ away from the steel surface since sodium chloride solution has a greater conductivity. On the other hand, $Fe(OH)_2$film adjacent to the steel surface can provide an effective diffusion-barrier film. It needs to be noted that chloride ions are usually the most sensitive media for both pitting corrosion and crevice corrosion (Hou 2019).

2.3.2 Material factors

2.3.2.1 Chemical composition

Although the composition of steel differs in different types of steel, mild steel, which is the mostly used structural steel, is primarily comprised of base material iron (Fe) and other five minor chemical elements, mainly carbon (C), manganese (Mn), silicon (Si), phosphorus (P) and sulphur (S). In a neutral environment, such as natural water or air, elements of steel within the design limits of the product have little or no significant effect on the corrosion rate of steel. This is mainly because the corrosion rate under such an environment is dependent on the diffusion of oxygen to the steel surface. In an acidic environment, however, the corrosion rate depends on the composition as well as the microstructure of steel and increases with both carbon and nitrogen content. Early research (e.g., Foroulis and Uhlig 1965) had shown that in an acidic environment, silicon hardly affects the corrosion rate of steel, whilst corrosion rate increases approximately linearly with the increase of contents of both phosphorus and sulphur in steel. Also, phosphorus affects the corrosion rate of steel more than sulphur. Recent studies also show that the corrosion rate of steel increases with the increase of its carbon content, especially in the range 0.5%–0.7%. This may not affect structural steel whose carbon content is between 0.1% and 0.25%. On the other hand, manganese generally improves the corrosion resistance of steel.

2.3.2.2 Microstructure

In addition to chemical composition of elements, the microstructure of steel refers to grain size, iron phases and distribution of impurities in the steel. In general, the microstructure of steel affects its resistance to corrosion (Marcus 2011). Usually smaller grain size in steel helps to maintain the stability and adherence of the passive oxide films formed on the steel surface before and during corrosion, which subsequently protect steel from corrosion and prevent further corrosion (Marcus 2011). Steel or iron contains two main phases in terms of its crystal structure, namely ferrite and pearlite. Ferrite is known as α-iron (α-Fe), and pearlite is composed of ferrite (α-Fe) and cementite (Fe_3C). In general, ferrite is corrosion prone, and cementite is corrosion resistant. The role of cementite (Fe_3C) can be more complicated. On one hand, a larger proportion of cementite in iron improves the stability of the passive oxide films (Ralston and Birbilis 2010), which has a positive effect on corrosion resistance of steel. On the other hand, cementite can promote corrosion after the passive oxide films are broken down when it forms a coherent network on the surface of steel. Impurities are mainly those residuals from the raw materials, such as iron ore and lime, that are left behind after the steel making process. In general, impurities in steel accelerate the corrosion reaction by creating stress concentration as well as galvanic reaction (Syugaev et al. 2008). Corrosion rate of steel can be increased with the presence of stress (see more in Section 5.2). Specific impurities in steel segregate grains at their boundaries, which can lead to intergranular corrosion (Revie and Uhlig 2008).

2.3.2.3 *Defects*

A lack of uniformity is one of the major causes of corrosion in steel, such as galvanic corrosion and pitting corrosion. Defects include imperfection on the surface of steel and bubbles, voids and dislocations at the microstructural level. Defects affect the electrochemical properties of the surface. Defects are starting points for corrosion since they can easily become anodic to uniform surfaces. Even edges or holes on the surface of steel can be anode for corrosion. At microstructural level, dislocation and lattice vacancies as well as interstitial atoms can increase the diffusion rate of specific impurities or alloyed components. This may affect corrosion. For example, hydrogen embrittlement occurs due to the accumulation of hydrogen at voids or defects, which subsequently leads to inner pressure increment. Although considerable research has been undertaken, it is still unclear how these defects affect the diffusion rate and subsequent corrosion.

2.3.3 Soil factors

Steel structures are also built underground in soil. Typical examples are underground pipelines which are essential infrastructures for a nation. Both steel and cast iron are widely used for pipes buried in soil. Strictly speaking, cast iron is a kind of steel, and thus, these two words are exchangeable in this section. Soil is a complex and dynamic system. Its chemical and physical properties change spatially and seasonally due to climates, human activities and plants. To understand corrosion progress in soil, it is necessary to thoroughly examine the effect of each soil property on corrosion behaviour.

2.3.3.1 *Water content*

The water content of soil is perhaps the most basic and important parameter for soil. It is widely believed to have a significant effect on corrosion of steel in the soil. Generally, corrosion rates of steel in soils with moderate moisture are higher than that in extremely dry or fully saturated soils (Wang et al 2018b). At low water content, steel is rapidly oxidised into a passage film that prohibits the diffusion of water and oxygen. High water content can cause the migration of ferrous ions from the steel surface to soil before being oxidised and accumulate on the surface. Water content can also promote corrosion reactions by reducing the resistivity of soil. In fully saturated soils, however, the corrosion process may cease since the water immerses the soil and steel, leading to a deficiency of oxygen supply (Kreysa and Schütze 2008). It is generally believed that there is a critical water content in soil with which a maximum corrosion rate of steel can be reached. This critical water content in soil is approximately 65% of its water-holding capacity (Wang 2018). However, not everyone believed that such a critical value was found (Murray and Moran 1989). This may be because, in the field, the water content in soil changes continually as it is a function of many factors, such as soil type, climate and geometric conditions.

2.3.3.2 *Soil resistivity*

As corrosion is an electrochemical reaction, soil resistivity plays a major role in determining the corrosion current. Soil resistivity is often used to evaluate the corrosivity of soils. There is a widely accepted qualitative relationship between soil resistivity and corrosivity, in which they are approximately inversely proportional (Roberge 2007). However, the effect of soil resistivity on corrosion behaviour of steel or iron has been subjected to debate in the research community. For example, Logan et al. (1937) found a very weak correlation between soil resistivity and pit depth. More recently, Petersen and Melchers (2012) found that resistivity of soil had an effect on the corrosion of macrocells, which are built over a long distance in the bulk of the soil; however, it had no effect on the corrosion of microcells, formed by the non-homogeneity of soil. In most cases, the analysis of the effect of soil resistivity on corrosion is often complicated, as other secondary factors (such as moisture, soil porosity, salt content and environmental temperature) greatly affect resistivity and usually interact.

2.3.3.3 *Soil pH*

The pH affects corrosion in almost all environments. The pH value of soil is known to affect corrosion reaction by acting as a reducing agent in the electrode reaction and influencing the corrosion cell potential (Marcus 2011). In general, the corrosion rate of buried steel increases dramatically when pH decreases from 4 to 3, whilst the corrosion rate does not appear to rely on pH when soil pH is over 5 (Kreysa and Schütze 2008). An empirical relationship showing the dependence of corrosion rate (r) on the concentration of hydrogen ion (C_{H^+}) can be presented as follows (Silverman 2003):

$$r = k(C_{H^+})^n \tag{2.7}$$

where k and n are constants. Although this empirical relationship can be observed in some solutions, the effect of pH on the corrosion of buried steel is complex and uncertain in most cases. In addition, the soil pH itself is affected by many variables, such as the content of carbon dioxide, organic acid, minerals and contamination by industry wastes (Kreysa and Schütze 2008). Generally, the corrosion process is slower in neutral or alkaline soils (pH from 5.5 to 8.5), except in the presence of microorganisms, such as sulphate-reducing bacteria (Doyle et al. 2003).

2.3.3.4 *Soil texture*

The texture of soil affects the corrosion of steel buried in it. This is mainly due to its influence on diffusion of gases and salts in soil to the surface of steel. The soil texture can also affect the movement of corrosion products, such as free expansion, which indirectly affects corrosion process of steel buried in the soil (Flitton and Escalante 2003). In general, soils with finely dispersed structures can increase corrosion rate of the buried steel because such soils retain moisture more easily than other soils. Soils with high moisture content not only significantly reduce the resistivity of soil but also promote the diffusion and migration of corrosion products outward into surrounding soil. Sandy soil affects the corrosion by increasing the aeration and movement of water and gases within the soil due to its large particle size. Clay soil also facilitates corrosion progress because it has a large content of dissolved ions in the pore water (Doyle et al. 2003). Furthermore, soils with a large content of clay and silt would increase the corrosion rate of buried steel since they are expected to shrink and crack during drought conditions, providing access for oxygen (Pritchard et al. 2013).

2.3.3.5 *Other factors in soil*

The temperature, sulphate-reducing organisms and astray current in soil are known to directly or indirectly promote the corrosion of buried steel (Wang et al 2018a). Since the role of temperature in corrosion in atmosphere is different to that in soil, it needs to be described in more detail. The temperature in soil can affect the soil resistivity, solubility of oxygen in soil pore water, oxidation reaction of steel and the property of protective oxide film. The corrosion rate of buried steel can be doubled if the temperature increases by 10°C. It should be noted that whilst an increase in temperature can increase the corrosion process, it can also result in evaporation and loss of moisture, slowing down the corrosion due to moisture loss. The presence of sulphate-reducing organisms in soil can accelerate corrosion in buried steel (Davis 2000). This is an area in which not sufficient knowledge, in particular, quantitative knowledge, is available. Some tests have been conducted, such as Wasim (2018), and clearly more need to be done. Stray current refers to the current that does not flow in the intended circuit or path. The corrosion of buried steel caused by stray current can be more serious than that by other soil factors. Buried steel has a high electrical conductivity and potential differences with the less conductive soil. A corrosion cell can be formed in the presence of stray current in soil. As such, the stray current accelerates the corrosion of steel buried in soil.

Whilst there are many factors that affect the corrosion of steel in soil, the analysis of these factors is often complicated by the interaction between the phases of solid, liquid and gas of the soil. Furthermore, most of these factors affect each other, and some of them (such as temperature and moisture) can impose opposite effects on the corrosivity of soil. As a result, it can be difficult to determine a single most significant factor that affects corrosion of steel or cast iron or in general ferrous metals in soil.

2.4 EFFECTS OF STEEL CORROSION

The effects of corrosion on steel or ferrous metal as a material can be in many forms, such as physical, chemical and microstructural. The most direct effect is the physical mass loss, due to rusting in the electrochemical reactions. This will reduce the geometry of the steel body, i.e., steel members. During the corrosion process, hydrogen is released which accumulates or is trapped inside steel. The trapped hydrogen reacts with other chemical elements and makes the steel brittle, the so-called hydrogen embrittlement. With continuous corrosion, the mechanical properties of steel may be affected which is the theme of this book and will be described in detail in next chapters. When corrosion goes deeper into the steel body, the microstructure of steel may be affected, including element composition, grain size, morphology and iron phase composition (Li 2018). This section focusses on the effect of corrosion on mass loss, hydrogen accumulation and microstructure, whilst the effects on mechanical properties are to be discussed in detail in next chapters.

2.4.1 Physical effect

The physical effect of corrosion on steel mainly refers to its loss of mass or geometric dimensions which is a quick and direct effect of corrosion. Both uniform corrosion and pitting corrosion can lead to loss of cross-section of the steel member. For uniform corrosion, there are several models to determine the loss of cross-section based on a known corrosion rate. Some standards are also available that provide guidance on corrosion rate for different environments. For example, Australian Standard AS 4312 (Australian Standard 2008b) classifies corrosive environments into five categories, as presented in Table 2.3, with different expected corrosion rates, respectively.

There are also many developed models that can predict the corrosion loss in geometry. The model of power law developed by Kayser and Nowak (1989) is perhaps the most widely used model for corrosion loss prediction. The power law function is expressed as follows:

$$C = kt^m \tag{2.8}$$

where C is the corrosion loss (thickness loss) in μm (micrometre) after the exposure time of t (year), k is corrosion loss when $t = 1$ and m is a regression constant. Both k and m are determined mainly by experiments.

The power law function can only predict the corrosion rate for steel exposed to atmosphere in a very short period (within 10 years) (Landolfo et al. 2010). For long-term exposure, a bi-linear law function was developed as follows:

$$C = C_r t, \text{ when } t \le 10 \tag{2.9a}$$

$$C = 10C_r + C_{r\text{lin}}\ (t - 10), \text{ when } t > 10 \tag{2.9b}$$

where C is the corrosion loss, C_r is the average corrosion rate (μm/year) in 10 years, $C_{r\text{lin}}$ is the steady state corrosion rate (μm/year) after 10 years and t is the time in years. The values of C_r and $C_{r\text{lin}}$ are determined according to ISO 9224 (International Organization for Standardization 2012).

Table 2.3 Corrosivity Categories

AS 1413 Category	*Corrosivity*	*Steel Corrosion rate (μm/year)*	*Typical Environment*
C1	Very low	<1.3	Dry indoors
C2	Low	1.3–25	Arid/urban inland
C3	Medium	25–50	Coastal or industrial
C4	High	50–80	Seashore (calm)
C5	Very high	80–100	Seashore (surf)

For pitting corrosion, the model to predict pit depth can also follow the power law function, bi-linear law function and indeed other models but the related parameters, such as k, m, C_r and $C_{\eta_{\text{lin}}}$, need to be determined from respective experiments.

The mass loss due to corrosion can be theoretically determined by Faraday's law as follows (Mangat and Molloy 1992):

$$m = \frac{MIt}{zF_a} \tag{2.10}$$

where m is the mass of steel consumed (in g), I is the current (in amps), t is the time (in s), F is the Faraday constant, which is equal to 96,500 C/mol, z is the ionic charge and M is the atomic weight of metal. For iron, $M = 56$ g, and $z = 2$. Thus, once the corrosion current is obtained through, such as linear polarisation resistance (LPR) measurement, the mass loss of steel due to corrosion can be determined.

The corrosion current can be measured by electrochemical methods, such as linear polarisation resistance. In this method, polarisation resistance R_p is experimentally determined and related to the corrosion current density through the Stern-Geary equation as follows (ASTM International 2004a):

$$i_{\text{corr}} = \frac{B}{R_p}, \text{where } B = \frac{\beta_A \beta_C}{2.303\,(\beta_A + \beta_C)} \tag{2.11}$$

where i_{corr} is the corrosion current density (μA/cm^2) and $i_{\text{corr}} = I\,/\,A$ with A being the surface area of corroding steel (in cm^2). The polarisation resistance R_p (Ω/cm^2) can be determined by potentiodynamic polarisation resistance measurement or stepwise potentiostatic polarisation measurement (ASTM International 2004a); B is the Stern-Geary constant, and β_C and β_A (μV/decade) are Tafel slopes either experimentally measured or estimated based on experience (Andrade and Alonso 1996). It should be noted that the accuracy of corrosion current measurement is subjected to debate amongst researchers and practitioners, in particular in field situations. This limits the widespread application of electrochemical methods to determine the mass loss of steel corrosion.

2.4.2 Chemical effect

The chemical effect of steel corrosion mainly refers to hydrogen production during the corrosion reactions, which is released and trapped inside the steel. With the progress of corrosion, the concentration of hydrogen increases which exerts local stresses and leads to the embrittlement of the material (Li et al 2018a). This is the well-known hydrogen embrittlement. Steel is one of the most susceptible metal to hydrogen embrittlement, which is a key mechanism for changes of microstructure of steel and hence its mechanical properties.

In electrochemical reactions, the cathodic reaction of steel corrosion can be rewritten as follows (Eggum 2013):

$$3Fe^{2+} + 4H_2O \rightarrow Fe_3O_4 + 8H^+ + 2e^- \tag{2.12}$$

from which process hydrogen is released. Cathodic hydrogen is adsorbed on the surface as atomic hydrogen (reduced). The accumulated hydrogen ions (H^+) can then ingress

Table 2.4 Location and Type of Hydrogen Trapping

Location	*Interstitial Holes*	*Lattice Vacancies*	*Dislocation*	*Voids*	*Impurities*	*Grain Boundaries*	*Phase Transitions*
Type	Weak	Weak	Reversible	Reversible	Reversible	Reversible	Reversible

to the surface of steel due to its surface energy. Afterwards, the hydrogen atoms can either form hydrogen molecules or diffuse into the steel body driven by the concentration gradient between the surface and the interior (Eggum 2013). Hydrogen, both as an atom and in gas phase, can be trapped within the steel (Chalaftris 2003, Revie and Uhlig 2008, Eggum 2013). Many locations within the steel can trap hydrogen, some of which are considered as reversible and some irreversible, as summarised in Table 2.4. Once hydrogen has been absorbed by a material, its effect, regardless of the source from where it has been absorbed, is the same.

Once hydrogen enters and is trapped within the steel, the mechanical properties of steel can be changed. The process by which the mechanical properties of steel are changed due to the introduction and subsequent diffusion of hydrogen into the metal is defined as hydrogen embrittlement (Chalaftris 2003). The mechanism of hydrogen embrittlement can be explained by internal pressure theory. In this theory, hydrogen embrittlement occurs due to the increase of concentration of hydrogen atoms trapped at various locations. The trapped hydrogen atoms can combine to form molecular hydrogen and create high pressure at the trapped site (Woodtli and Kieselbach 2000, Chalaftris 2003). The high pressure initiates cracks and degrades the ductility of steel.

Hydrogen embrittlement occurs during the plastic deformation of steel in contact with hydrogen gas and is strain rate-dependent. Hydrogen embrittlement results in a brittle fracture throughout the embrittled material as a result of hydrogen adsorption unless the strength of the remaining material is less than the load applied. Later, instantaneous final fracture occurs. The failure by hydrogen embrittlement is mostly intergranular. The fractured surface has, therefore, a crystalline appearance.

The mechanism of hydrogen embrittlement has not been definitively understood, sufficing to identify hydrogen as a cause for cracking. A widely held view is that impurity segregations at the grain boundary act as agents and increase the adsorption of cathodic hydrogen at these sites. It is also believed that hydrogen embrittlement is triggered by the interaction of hydrogen with defects in the metal, such as voids, dislocations, grain boundaries and so on. Hydrogen is trapped in these defects and facilitates the growth of a crack. A large number of such defects interact with hydrogen, and the combined trapping results in a significant loss of ductility. More details of hydrogen embrittlement and its impact on mechanical properties of steel will be discussed in Section 5.5.

2.4.3 Microstructural effect

Microstructural effect of corrosion on steel refers to changes in element composition, grain size, iron phase composition and morphology of the microstructure (Horner et al. 2011, Gonzaga 2013, Zhou and Yan 2016). The composition of chemical elements of mild steel can affect its mechanical properties (Li et al 2018a, Lino et al. 2017). The major chemical elements of mild steel include iron (Fe), carbon (C), manganese (Mn), silicon (Si), phosphorus (P), sulphur (S), aluminium (Al) and chromium (Cr). During corrosion, the electrochemical reactions (Equations 2.1–2.6) can reduce the content of iron in steel (De la Fuente et al. 2011). Carbon content may be subsequently increased since they are not reacting with acid. The proportions of other alloying elements, such as manganese, phosphorus, silicon, aluminium and chromium, can be reduced during corrosion since they can either be washed away or reacted with the corrosive solution (Revie and Uhlig 2008, Zhou 2010, Eggum 2013).

Other elements will ingress into the steel during the process of its corrosion. This will change the composition of the steel. Two main elements that ingress into steel during corrosion are oxygen and chloride (Revie and Uhlig 2008). The oxygen content increases due to the formation of brittle rust layers during the corrosion process. Chloride content increases because chloride ions break the passive oxide film formed on the steel surface which makes the steel vulnerable to pitting corrosion and stress corrosion cracking (SCC). Steel in a chloride-enriched environment is prone to chloride penetrations.

Corrosion affects the grain of steel by reducing its size due to intergranular corrosion. Intergranular corrosion, by definition, is the preferred corrosion at grain boundaries (Sinyavskij et al. 2004, Zhou and Yan 2016). This type of corrosion occurs because the boundaries of grains are more susceptible to corrosion than their centres, as the alloying elements in steel are likely to be depleted at the grain boundaries. Intergranular corrosion weakens the bonding force between grains and makes grain boundaries vulnerable to cracking. The interaction of stress and corrosion can further reduce the grain size by causing SCC along grain boundaries, i.e., initiation and growth of intergranular SCC (IGSCC) (Revie and Uhlig 2008, Marcus 2011). IGSCC is the initiation and growth of cracks by localised corrosion along the grain boundaries in steel with the presence of stress.

For the corrosion effect on phase composition, steel contains two main phases of iron judging from its crystal structure – ferrite and pearlite. As discussed in Section 2.3.2, ferrite is the body-centred cubic (bcc) structure of iron, known as α-iron (α-Fe), which provides ductility of steel, and pearlite is composed of ferrite (α-Fe) (87.5% by weight) and cementite (Fe_3C) (12.5% by weight), which makes steel brittle (Gonzaga 2013). Ferrite is corrosion prone since it is extracted from iron ore with high free energy. Cementite is corrosion-resistant (Sun et al. 2014). Thus, it is expected that corrosion can change the proportion of ferrite and pearlite in steel. Also, as pearlite is more brittle than ferrite, the composition of pearlite can be reduced in combined stress and corrosion environment due to the pearlite fracture (Gonzaga 2013). Corrosion affects the morphology of microstructure of steel through initiating pits and cracks.

2.5 CORROSION CHARACTERISTICS OF FERROUS METALS

There are many types of steel, some of which have been described in Section 2.1 and most of which can be found in other specialist books on steel types (Revie and Uhlig 2008). Different types of steel in this section and indeed in this book refer specifically to low carbon or mild steel (or structural steel), cast iron and ductile iron, which are collectively known as ferrous metals. By definition of steel as an iron base alloy, all these three metals can be called steel. It is known that these three metals behave sharply differently in mechanical terms. For example, steel is elastoplastic, cast iron is very brittle and ductile iron is in between. It is also known that the corrosion behaviour of these three metals is quite different, but exactly how and why they are different is less known. This is one of the purposes of the present book.

2.5.1 Difference in material

The difference in these three metals is primarily in material compositions. In general terms, steel is ferrous alloy consisting mostly of iron (Fe) as its base material with other minor elements. The difference in steel, cast iron and ductile lies mainly in their compositions of chemical elements and iron phases. Steel is more iron-based alloy, containing typically less than 1% carbon in content. Mild steel or structural steel typically contains 0.15%–0.20% carbon. Cast iron and ductile iron are also iron-based alloy but contain usually more than 2% carbon in content. The typical range of chemical composition for these three types of steel is summarised in Table 2.5 (Li 20018, Wasim 2018).

Due to larger proportion of carbon contents, cast iron contains perhaps more graphite which is a crystalline allotropic form of carbon. The shape and size of the graphite are dependent on the process of steel making. A slow cooling process leads to large graphite flakes, and fast cooling results in fine graphite (Bradley and Srinivasan 1990). During solidification, the major proportion of the carbon precipitates in the form of graphite. Ductile iron is a type of graphite-rich cast iron. It is also known as ductile cast iron or nodular cast iron. The key difference in cast iron and ductile iron is the shape of graphite, namely, the graphite in ductile iron has a nodular or spheroidal shape. During the manufacture of ductile iron, magnesium is added, which causes the carbon in the metal melt to precipitate upon solidification in the form of graphite nodules within the ferritic alloy matrix. The spheroidal shape of graphite in ductile iron makes it more ductile than cast iron whose graphite is of flaky shape. The reason may be that there is less stress concentration at the boundaries of graphite with spheroidal shape than that with flaky shape.

There are two main phases of iron involved in steel or iron alloy which are ferrite and pearlite as discussed in Section 2.3.2. It should be noted that iron phases are not the same as structures of iron or steel. In comparison, steel contains mostly ferrite, a form of pure iron (α–Fe) with a body-centred cubic crystal structure, and small proportion of cementite, whilst cast iron and ductile iron contain relatively more cementite (Fe_3C)

Table 2.5 Chemical Composition of Three Selected Steels (wt %)

Type of Steel	*C*	*Mn*	*Si*	*P*	*S*
Mild steel	0.15–0.2	0.08–1.50	0.40–0.80	<0.03	<0.05
Cast iron	2.5–4.0	0.2–1.0	1.0–3.0	0.02–1.1	0.02–0.25
Ductile iron	3.0–4.0	0.1–1.0	1.8–2.8	0.01–0.1	0.01–0.03

than steel. There is no definitive proportion of each phase in each steel, not in quantitative terms. Heat treatment can change the proportion of each phase in the steel.

2.5.2 Difference in corrosion

As described in Equation (2.1), steel corrosion is an electrochemical reaction of iron (Fe) with surrounding environments. Iron contains two phases as discussed in Section 2.5.1. It is in the ferrite phase that corrosion occurs for all three steels, i.e., steel, cast iron and ductile iron, which involves the same electrochemical reactions. Based on this, it is fair to say that the basic corrosion behaviour of these three steels is by and large the same. However, there is a difference in their corrosion behaviour mainly due to different carbon contents. For cast iron and ductile iron, graphite exists in the iron matrix formed during the process of steel making (Bradley and Srinivasan 1990). Thus, in addition to electrochemical corrosion, there is a graphitic corrosion, in which the metallic constituents are leached out or turned into corrosion products, leaving the graphite intact and exposed. Graphitic corrosion is often seen in cast iron pipes buried in soil. It is a serious form of deterioration of cast iron pipes. Graphitic corrosion occurs after corrosion in ferrite phase because of their different corrosion activations (Romanoff 1957).

With the existence of graphite, corrosion of cast iron occurs preferably along the boundaries of graphite flakes, resulting in deep pitting corrosion. Corrosion of ductile iron is less deep than that of cast iron since ductile iron has dispersed graphite nodules, but it is still deeper than that of steel. However, the difference in the pitting is not significant between cast iron and ductile iron in the same environments as observed from the field burial tests (Romanoff 1964). Further, it appears that the exposure environment has more influence on the corrosion behaviour of cast iron and ductile iron than the variations in their materials (Kreysa and Schütze 2008). The difference in corrosion between steel, cast iron and ductile iron is mainly in that steel tends to be prone to more uniform corrosion, whilst cast iron and ductile iron are prone to pitting corrosion, in particular, cast iron in a complex environment such as soil.

Since the environment affects corrosion significantly, the difference in corrosion amongst these steels can be different for different environments. In other words, in one environment, steel may suffer more corrosion than cast iron and ductile iron, and in another environment, vice versa. Also, in one environment, steel may exhibit uniform corrosion and cast iron may exhibit pitting corrosion, but in another environment, vice versa. This just indicates the nature and complexity of steel corrosion and explains why there is so much literature and research on steel corrosion, and yet more will come.

There are different views on corrosion behaviour of ferrous metals, i.e., steel, cast iron and ductile iron. In general, steel is more corrosion-resistant than cast iron in the same corrosive environment. Cast iron is least corrosion-resistant amongst these three steels, whilst ductile iron is most corrosion-resistant amongst these three steels, but the difference in corrosion resistance is not significant in a neutral or atmospheric environment.

2.5.3 Comparison of corrosion

It would be of great interest, practically and academically, to quantitatively compare the differences in corrosion of three ferrous metals. This is possible with data produced

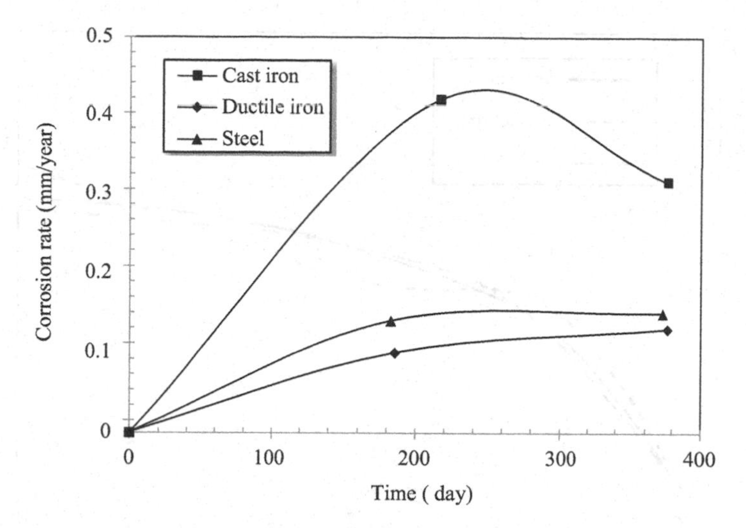

Figure 2.5 Comparison of corrosion of ferrous metals in acidic solution.

from laboratory tests and collected from field inspection and tests. Figure 2.5 shows the laboratory test results of corrosion of the three ferrous metals, i.e., steel, cast iron and ductile iron in the same acidic solutions for various periods of immersions. It is clear from the figure that cast iron corrodes more and faster than both steel and ductile iron in the same corrosive environment. In particular, as it can be seen, in the middle of the corrosion process, the corrosion rate of cast iron is about three and four times higher than that of steel and ductile iron, respectively. In the longer term, this difference reduces to two times for both steel and ductile iron. In general, cast iron corroded most, steel second and ductile least in this environment. The reason for more corrosion of cast iron can be mainly due to surface morphology of the cast iron, which is rougher and less uniform than both steel and ductile. Rough surface facilitates easier ingress of corrosive agents, in particular, dissolved oxygen. In the longer term, however, corrosion may have reached a certain stage that the corrosion products, i.e., rusts, cover the surface which makes such agents difficult to ingress to the surface to cause further corrosion.

Of many types of corrosion, uniform corrosion and pitting corrosion are of most practical importance to all stakeholders of steel producers and steel structures. It would be of significance to compare how three different ferrous metals corrode in the same environment. Laboratory tests are difficult to produce pitting corrosion since all specimens are immersed in the solution which is more or less uninform. In this case, it is more appropriate to collect data from field tests. A comprehensive data mining has yielded sufficient data on both uniform corrosion and pitting in soil (Romanoff 1964). Figure 2.6 shows the comparison of corrosion of three ferrous metals, i.e., mild steel, cast iron and ductile iron.

It can be seen from Figure 2.6 that for uniform corrosion, there is little difference amongst steel, cast iron and ductile iron. The maximum difference between any of the metals is less than 8.86%. This may be understandable since all three metals undergo the same electrochemical reactions at macro-scale largely in the ferrite phase. On the

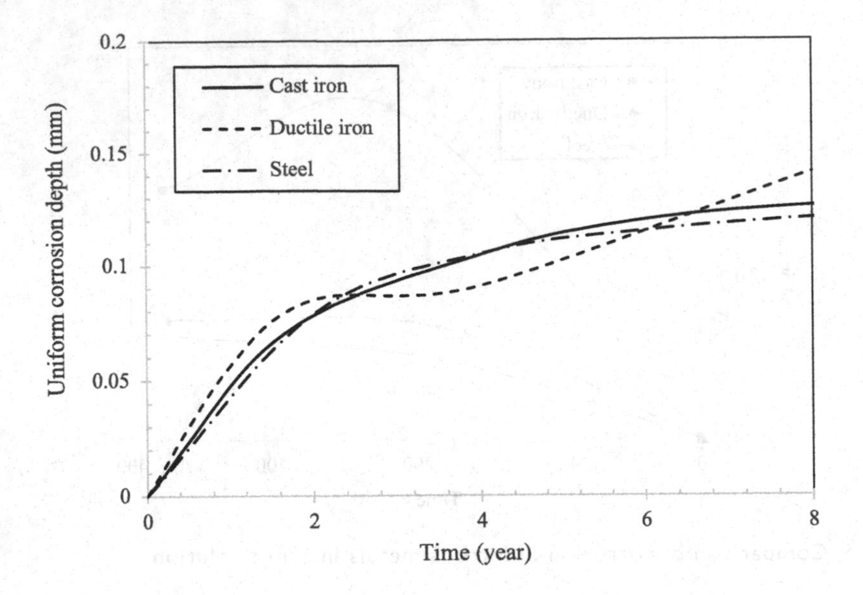

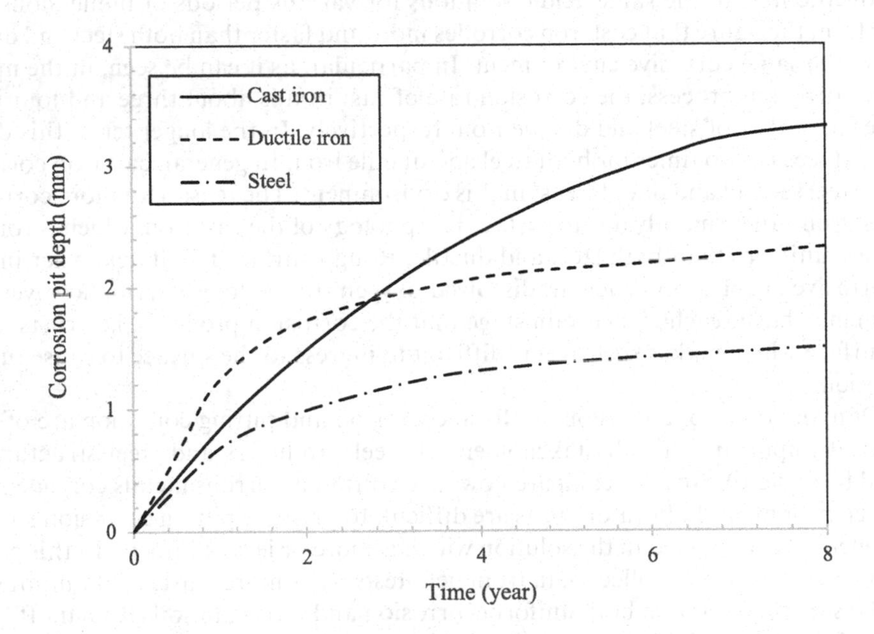

Figure 2.6 Comparison of corrosion of ferrous metals in soil.

other hand, the pitting corrosion is quite different for different steels. Cast iron suffers the highest pitting corrosion with steel the least. This again can be explained from their different chemical compositions, in particular, graphite. With more graphite in cast iron, corrosion tends to be localised along the edges of flaky graphite which leads to deeper corrosion, i.e., pitting. In addition, graphitic corrosion further increases the pitting corrosion. It is of interest to note that the difference in corrosion is different for different steels and becomes larger with longer exposure time. For example, at year 1, ductile iron has the highest pitting corrosion with steel being the least. However, at year 8, cast iron has the highest pitting corrosion again with steel being the least. At year 1, the largest difference in pitting corrosion is between ductile iron and steel, which is about 77%; whilst at year 8, the largest difference in pitting corrosion is between cast iron and steel, which is as high as 110%. It is hoped that this comparison can be of great interest to practitioners and researchers alike in assessing corrosion of different ferrous metals.

2.6 SUMMARY

To understand the corrosion of steel more profoundly, it is necessary to know how steel is made, which differentiates different types of steel and in particular, mild steel, cast iron and ductile iron. It is also necessary to know that the chemical composition of steel affects the corrosion and the mechanical properties of steel. Following this, the detailed process of steel corrosion, i.e., the electrochemical reactions, is presented together with a conceptual model of corrosion progress. Types of corrosion are also discussed in Section 2.2 with more focus on practically significant forms of corrosion, such as uniform corrosion and pitting corrosion. The factors that affect the steel corrosion in different environments are discussed in Section 2.3. Whilst it is known that there is no such a single most significant factor that affects corrosion of steel, it is generally accepted that the environment will affect the corrosion of steel the most. Section 2.4 covers the effects of corrosion on steel from a material perspective. Knowing that the effect of steel corrosion on mechanical properties of steel will be covered in length in later chapters, this section only discusses corrosion loss, hydrogen concentration and changes in microstructure of steel. Finally, a comparison of corrosion of steel, cast iron and ductile iron is presented in Section 2.5 both qualitatively and quantitatively with both laboratory test data and filed data. It is shown that in the same environment, steel corrodes the least whilst cast iron the most in particular for pitting corrosion. This comparison can be of great interest to practitioners and researchers alike.

Chapter 3

Corrosion impact on mechanical properties of steel

3.1 INTRODUCTION

Corrosion of steel has been widely recognised as the main cause for structural deterioration and ultimate collapse of corroded steel structures. It reduces the expected design life of steel structures. Therefore, it is imperative to accurately predict corrosion impact on structural steel, in particular the impact on degradation of mechanical properties of corroded steel in order to ensure the safe and reliable operation of steel structures during their expected service life.

In most of the current practice for design and assessment of steel structures, corrosion effect is accounted for by considering the loss of cross-sectional area of steel or steel members over the design life span. If this is adequate, the likelihood of structural failures, in particular, the collapse of such designed structures, should be very small. However, as evidenced in Chapter 1, the likelihood of structural failures of corrosion-affected steel structures cannot be considered as small. Thus, either the corrosion loss has not been considered adequately due to more severe corrosion, or there may be some other factors that have not been considered in the design and assessment of corrosion-prone steel structures. Either way, there is a need to continue to investigate the corrosion of steel and its effect on degradation of mechanical properties of corroded steel. The hypothesis adopted here is that the corrosion not only reduces the cross-sectional areas of steel members but also reduces the mechanical strength or properties in general of the corroded steel, such as tensile strength, fatigue strength and fracture toughness. This hypothesis is supported by considerable evidence recently published by many researchers, such as Revie and Uhlig (2008), Marcus (2011), Eggum (2013) and Li (2018), to name a few.

This chapter presents quantitative results of corrosion impact on degradation of mechanical properties of steel, with focus on tensile properties (i.e., yield strength, ultimate strength, etc.), fatigue strength and fracture toughness. These properties are the most important parameters of design and assessment of steel structures. In this chapter, how to conduct corrosion test with a view to investigate its impact on mechanical properties is presented first. This includes design of test specimens, test procedure and measurement of corrosion parameters. Data on the corrosion loss of steel members of a structure exposed in natural atmosphere are also presented. Then degradation of tensile properties due to corrosion is fully examined, including yield strength, ultimate strength and failure strain. The mechanisms behind this degradation are also discussed including changes in the microstructure of the corroded steel. After that, data of corrosion effect on degradation of fatigue strength and fracture toughness of steel are presented, and mechanisms of this degradation are analysed.

The approach that this chapter and indeed the whole book adopts is a mechanistic one based on phenomenological observations from laboratory experiments and field inspections. It is believed this is a realistic and effective approach in dealing with corrosion effect on mechanical properties of corroded steel without losing the basic rigour of the corrosion process, i.e., the electrochemical reactions. This approach can produce results that are directly related to steel and steel structures to be designed and assessed. Although the approach is more based on phenomenological observations of laboratory tests and field measurements, the mechanism for degradation of mechanical properties of steel due to corrosion is discussed in detail and at both macro and micro levels. As discussed in Chapter 2, elemental composition affects the mechanical properties of steel. Intuitively, any changes in elemental composition in steel might affect the mechanical properties of steel. Other microstructural features, such as grain size and iron phase composition, are also possible factors that affect the mechanical properties of steel. Changes in the contents of these factors are possible causes for degradation of the mechanical properties of corroded steel and hence are discussed in the chapter.

3.2 OBSERVATION OF CORROSION

Natural corrosion, i.e., corrosion in natural environments, such as in atmosphere, marine or soil, may take years if not decades to manifest its impact on steel, either as the significant mass loss of steel material or as degradation of the mechanical properties of steel. For this reason and considering the working life of ordinary researchers and practitioners, acceleration is usually a viable alternative for studies on corrosion even though sometimes it is regarded as the last resort. This is particularly necessary when the purpose of corrosion tests is to determine its effect on the mechanical properties of steel since it takes even longer time for corrosion to cause any degradation of mechanical properties of the corroded steel. From the outset of this chapter and the subsequent chapters, it needs to be noted that the effect of corrosion on mechanical properties can be different under different accelerated corrosions. Calibration of results from accelerated corrosion tests against those under natural corrosion is an appropriate avenue to overcome this concern. Another approach is to examine the relative effect of corrosion on mechanical properties of steel for a given condition. This is the approach taken in the book and indeed by many of those studies that have been published in the literature, such as Li et al. (2018a, b), Wasim et al. (2019a, b) and Wang et al. (2018a, b), to name a few. This approach does provide useful information on corrosion-induced degradation of mechanical properties of steel. Discussions on the applicability of results obtained to design and assessment of practical steel structures are provided in Chapter 6. Having said that, corrosion in natural environments is also covered in this chapter although such cases or examples are not as many and as comprehensive as one would wish.

3.2.1 Simulated corrosion

The rationale to simulate corrosion, adopted in this chapter and indeed this book, is to select an appropriate corrosive environment that can generate sufficient corrosion effect on degradation of mechanical properties of selected steel, such as structural steel in this chapter. Since this chapter focuses on three identified mechanical properties of steel, namely, tensile strength, fatigue strength and fracture toughness, it is reasonable to directly use the mechanical test specimens for these properties respectively in the corrosion tests although this is not essential and sometime circumstances may not permit, such as space limit. To observe and measure the degradation of these mechanical preparties caused by corrosion, mechanical tests need to be carried out at a minimum of three points in time to develop any patten or trend of degradation over time. Of course, the more points in time there are, the more accurate the pattern or trend can be, but the reality is always that limited time and resources are provided or available for such requirements. Other important aspects of corrosion tests include the cleaning of corroded steel, such as rust removal, to ensure no damage to bulk steel would occur in such process.

To ensure the quality and credibility of the tests and in particular the data produced from the corrosion tests and subsequent mechanical tests, relevant test standards should be followed in conducting such tests presented in the chapter, including the procedure to clean the specimens after corrosion and preparation for mechanical tests. Details of such standards and importantly the procedures included in the standards are beyond the scope of the book but can be easily found in the literature. The following are a few examples:

- ASTM International. (2004b). ASTM G31-72, Standard practice for laboratory immersion corrosion testing of metals, ASTM International, West Conshohocken, PA
- ASTM International. (2016). ASTM E8/E8M-16a, Standard test methods for tensile testing of metallic materials, ASTM International, West Conshohocken, PA.
- BS 7608 (2014). Guide to fatigue design and assessment of steel products, British Standards Institution (BSI), London.
- ASTM International. (2018). ASTM E1820-18, Standard test method for measurement of fracture toughness, ASTM International, West Conshohocken, PA.

3.2.1.1 Exposure environments

Acidic solution is usually selected as the exposure environment for corrosion immersion test. This is mainly because the acidic environment can accelerate the corrosion. Two types of acidic solutions are discussed in this chapter. One is the generally used acidic solution, and the other is the specifically used solution simulating soil environment. Amongst many acidic solutions, hydrochloric acid (HCl) is often selected as the immersion solution (Noor and Al-Moubaraki, 2008). Selection of HCl solutions is better to cover a range of acidities (measured by the content of HCl in pH value or more precisely in molar) and also related to practical situations, at least some of them. The range of pH values of HCl solutions usually used in immersion tests varies from pH = 0 (1 M HCl) to pH = 5 (0.00001 M HCl). This range is wide enough to cover various corrosion conditions. Also, this range of corrosive environments can be encountered in the real world. For example, the solution with pH = 5 can represent natural environment where there is a large amount of organic substances, such as steel buried in soil

Table 3.1 A Sample of Chemical Composition in Soil Solutions (g/L)

Chemical	$CaCl_2{\cdot}2H_2O$	$MgSO_4{\cdot}7H_2O$	KCl	$NaHCO_3$
Content	0.036	0.190	0.069	0.540

(Liu et al. 2014a). Even the most acidic solution, i.e., pH = 0, can simulate steel wells in oil fields subjected to chemical cleaning (Finšgar and Jackson 2014).

Another acidic solution is for simulating the soil environment. The simulated soil solutions can be made by taking key elements from real soils, i.e., using the chemical properties of the key components of soil (Hou et al. 2016). Based on the principle that the key chemical elements of the soil sample and soil solution are the same (Liu et al. 2009), a simulated soil solution can be made. Table 3.1 is one of such simulated soil solutions with key elemental composition of the soil to be simulated (Hou et al. 2016). Again, three values of pH are selected to represent different acidities of the soil and its effect on corrosion, which can be determined based on research experience and pre-trials. For example, it is known that a pH of 3.0 can accelerate the corrosion to significant effect on the mechanical properties of the steel within a designated period of time. The pH of natural soil is about 8.0, so a middle value of pH = 5.5 can be selected to have a different effect of pH on degradation of mechanical properties. Different values of pH can be achieved and maintained by adding sulfuric acid (H_2SO_4) to the solution during the tests.

3.2.1.2 Test specimens

Materials used in corrosion test are usually low-carbon steel, i.e., mild steel or structural steel. There are various grades of such steel, and each country has its own standard for the manufacture and specification of the steel products. For example, G250 is a widely used structural steel in Australia, Fe430 in UK and Q235 in China. The chemical composition of such steels can be available from the supplier. A typical composition of mild steel G250 is shown in Table 2.5 of Chapter 2.

Test specimens are designed such that the subsequent mechanical tests can be carried out directly at the designated time of corrosion exposure. In this way, the effect of processing and manufacturing of the specimens after corrosion can be eliminated. Whilst the material of the specimens is the same for all corrosion and mechanical tests, the configuration of the specimens for different mechanical testing is different. To account for the variability of the test process and test results, it is appropriate to have a number of identical test specimens, i.e., duplicates. Statically, the minimum number of specimens that can take into account the variation of test results is 3. This is the number of identical test specimens presented in the chapter for a given corrosion condition and given mechanical test. In other words, most data points presented in the chapter represent an average of three measurements.

For the purpose of tensile test, the design specimen should follow a standard, such as ASTM E8/E8M (ASTM 2016). Different standards may specify different dimensions and configurations, but they are more or less the same or at least similar. For example, by American standard ASTM, specimens are dog bone-shaped prism, and by Chinese standard, the specimens are cylindrical. To ensure that the actual exposure area of the specimens is confined to only the expected corrosion area, such as the middle part of 50mm gauge length, it is a usual practice to insulate both ends of the test specimens by, e.g., wrapping them with acid-resistant tape. For the purpose of fatigue test, if the specimen design follows ASTM E466-15 (2015) *Standard practice for conducting force controlled constant amplitude axial fatigue tests of metallic materials*, the configuration of the specimens is similar to that for tensile test with a small difference in dimensions (Li 2018).

For the purpose of fracture toughness tests, the configurations are quire complicated compared with those of tensile and fatigue tests. The key dimension to be determined first is the width of the specimens, which determines plane strain or plane stress fracture condition. In addition to dimension, a notch needs to be cut at the middle of test specimens under three point bending for single-edge notched bending (SENB). Also, there are three modes of fracture, each specimen of which is different. Most current studies on corrosion-induced degradation of fracture toughness are on Mode I fracture, which is the fracture mode used in this chapter for fracture toughness test. Details of fracture toughness tests can be referred to in books such as Anderson (2017) and papers, including a recent review by Wang et al. (2020).

3.2.1.3 Test procedure

Corrosion simulation is achieved by immersing test specimens in the selected simulated solution environment. This is commonly known as corrosion immersion test. Firstly, the acidic solutions are made in an anti-acid container with designated pH values. Then, the specimens are placed in the solution as schematically shown in Figure 3.1. From the time the specimens are immersed in the solution, the corrosion tests start. Corrosion activities are monitored by the measurement of corrosion current, using linear polarisation resistance, which is taken continuously in the first few days. A week later, it is taken daily, and then after a few weeks, it is taken weekly depending on the duration of the exposure time.

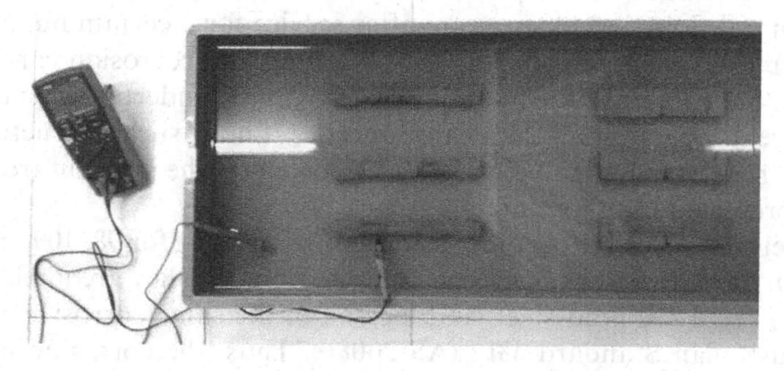

Figure 3.1 Immersion tests and monitoring of corrosion activity.

The exposure duration depends on the purpose of the tests and perhaps in most cases is determined by the project timeline. In principle, the longer the period of exposure, the more adequate the information obtained. In the published literature, the exposure time for accelerated corrosion can be as short as days, such as Li et al. (2018a) and as long as years, such as Wasim et al. (2019b). At a designated time of exposure, specimens are removed from the container for measurement of corrosion loss and then mechanical testing. After removal from the exposure environment, the corroded specimens need to be cleaned, again in accordance with a specific standard to ensure that it is well cleaned and also no undue damage occurs during the cleaning. For the example of ASTM G1-03 (2017), specimens should firstly be rinsed in rust removal solution and then grounded with superfine sandpaper (600 grit) to remove rust.

For the quality and reliability of test data, a minimum number of specimens in immersion tests and mechanical tests are required which varies and depends on the resources and time. It can be designed as follows: the number of solutions selected (usually three) × number of exposure periods (minimum three) × number of duplicates (minimum three), plus three reference specimens for no corrosion, which amounts to 30. This is a minimum required number to conduct meaningful corrosion tests in order to develop trends of corrosion effects on degradation of the mechanical properties of the corroded steel.

3.2.2 Natural corrosion

Simulated corrosion requires certain degree of acceleration which may alter or introduce some unexpected factors. Thus results produced from such tests require some kind of calibration before they can be applied to practical steel structures. Such calibration is often extremely difficult due simply to the fact that there is no method widely accepted in the research community and general practitioners for the calibration. The primary reason for lacking such an important method for calibration can be that it is often extremely difficult to have two exactly identical environments in which corrosion takes place. Also, corrosion in natural environment, such as atmosphere or marine, involves too many factors in the process of corrosion compared with corrosion in laboratories, and some of these factors cannot be identified nor controlled. Furthermore, natural corrosion takes much longer time, such as over 30 years, to produce meaningful results, in particular for corrosion effect on mechanical properties of the corroded steel. This makes the repeated tests (as required for all scientific tests) almost practically impossible.

One possible solution to the problem of calibration of accelerated test results is to find a decommissioned steel structure after service for a certain number of years. Then, steel samples can be taken from the structure, and corrosion can be observed and measured. Subsequently, the mechanical tests can be undertaken on the corroded steel to observe its effect on mechanical properties. Lucky such a structure is found, and necessary measurement and tests are undertaken on the steel cut from the structure as to be presented in this section.

Three steel structures are decommissioned after serving for 98, 109 and 128 years, respectively, in the natural environment, i.e., atmosphere. They are made of the same grade of steel, i.e., G250 mild steel, and located in the same corrosive zone as categorised by Australian Standard 4312 (AS 2008b). Thus, the corrosion measurement and subsequent mechanical tests on steel samples taken from any of these three steel structures can be compared and represent the corrosion loss and residual mechanical properties at their respective times.

General inspections are firstly carried out on these decommissioned structures to have an overview of the corrosion conditions. Based on the inspection, the corrosion condition is classified in three levels: mild corrosion with less than 1 mm thickness loss; moderate corrosion with 2 mm thickness loss and severe corrosion with more than 3 mm thickness loss. The thickness of the specimens is measured by an ultrasonic thickness device. Samples are then taken from each level of corrosion condition. They are later processed to make sufficient number of specimens for the planned tests, including strength tests, fatigue tests and microstructural tests.

All cut samples are rinsed in rust removal solution and then grounded with superfine sandpaper (600 grit) to remove rust. They are then cleaned thoroughly with bi-distilled water followed by acetone and dried with air. At each corrosion level, samples are processed to make three specimens for tensile tests. The dimensions of the specimens are the same as those for simulated corrosion tests, but the thickness of specimens is based on the actual thickness of the girder plates of the steel structure, which are 10, 12 and 15 mm, respectively, for three different structures.

Likewise, at each corrosion level, samples are processed to make three specimens for fatigue tests. The dimensions of the specimens cut from the decommissioned steel structure for fatigue tests should be in compliance with a standard. Again, ASTM E466-15 (2015) is widely used.

In addition to steel girders, three exhumed steel pipes are collected from water utilities. The pipes were used for water distribution at around 50 m of water head. The pipes were buried in soil with the burial depth from 0.8 to 1.5 m. The ages of the exhumed pipes and their nominal thickness are shown in Table 3.2. To determine the types of steel of these pipes, drillings from the undamaged areas (substrate) of pipes are analysed by a Varian 730-ES Optical Emission Spectrometer. The results of elemental compositions for the pipes are presented in Table 3.2, from which it can be seen that steel has a carbon content of less than 0.12% (by weight). This is the low carbon steel or structural steel as discussed in the chapter. Samples are then cut from the pipes to make specimens for other mechanical tests. All test specimens are carefully prepared so as to be representative of the corroded pipe.

3.2.3 Corrosion measurement

Due to the nature of the immersion test, it is assumed that the time the specimens are placed in the solution is the time corrosion starts. In other words, corrosion initiation

Table 3.2 Information of Pipe Samples and Element Composition Results

Pipe No.	*Age*	*Nominal Thickness (mm)*	*Element Composition (wt %)*				
			C	*Si*	*Mn*	*P*	*S*
1	52	5	0.12	0.03	0.35	0.02	0.02
2	55	4.8	0.07	0.03	0.42	0.02	0.02
3	63	5	0.10	0.09	0.41	0.04	0.04

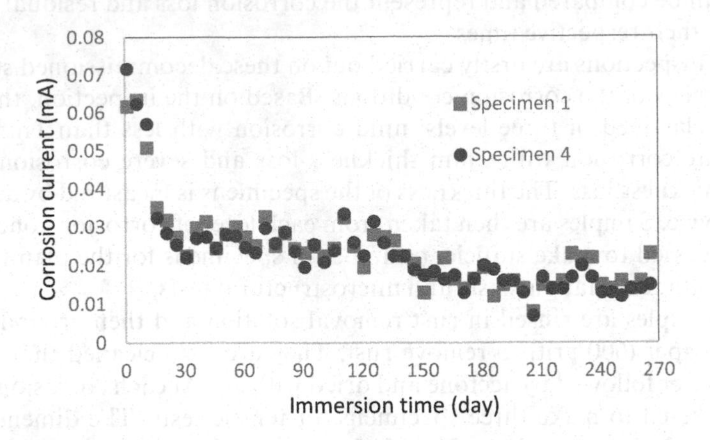

Figure 3.2 Corrosion current in specimens in acidic solutions.

is assumed instant and not considered in detail in this chapter. The corrosion progress can be monitored by corrosion current. Although the accuracy of this measurement is not without controversy, it can provide indication of corrosion activities. In general, higher corrosion current indicates more corrosion activities. One example of corrosion current measurement of steel specimens in acidic solution with pH = 3 is shown in Figure 3.2. As can be seen, at the onset of corrosion, the current is very large indicating a high potential at the anode. This makes sense since corrosion starts when there is a high potential between two points, which is the condition for corrosion to initiate. High corrosion current indicates more corrosion activities or high corrosion rate in terms of electronic current. After corrosion is initiated, the corrosion tends to progress steadily in which the corrosion current is more or less constant. Once corrosion progresses for a certain period of time, such as 230 days in Figure 3.2, corrosion tends to slow down, as indicated by smaller corrosion current, which indicates smaller corrosion rate. The primary reason for corrosion to slow down at this stage is that the corrosion products, i.e., rusts, accumulated at the anodes so that the diffusion of oxygen into the steel surface becomes more difficult or slower. The three stages of corrosion progress of Figure 3.2 are in line with the conceptual model of Figure 2.3 in Chapter 2 when it is measured by corrosion loss in mm.

Figure 3.2 indicates that, although each point of measured corrosion current is scattered, the general trend of corrosion currents is clear, which is decreasing with the exposure time. This means that the current rate is high at the beginning of the corrosion and decreases over time. As is well known, corrosion is an electrochemical process. The acidic environment can initiate the corrosion, but the progress of corrosion needs the supply of oxygen, which is not readily available to keep the high corrosion rate. These results are consistent with other results reported in the literature as well as research experience (Mohebbi and Li 2011).

Corrosion loss can be measured physically at each designated time. This is the most accurate and reliable measurement of corrosion progress compared with other means, such as corrosion current (Revie and Uhlig 2008, Wang et al. 2014). After a certain period of exposure time, such as 7, 14 and 28 days for a short-term project (Li 2018) or 30, 90 and 180 days for a longer-term project (Hou et al. 2016), three duplicate specimens are taken out of the container, i.e., immersion solution, and the tested part of the specimens is cut off for weighing. The tested areas should be washed thoroughly with bi-distilled water followed by acetone and dried with air to remove rust and stop corrosion.

The corrosion loss can be expressed in mass loss or thickness loss as discussed in Chapter 2. The mass loss (Δ) of the exposed part of the specimen at a given exposure time can be determined simply as follows:

$$\Delta m = m_0 - m_1 \tag{3.1}$$

where m is the average mass of the exposed part (tested area) of three duplicate specimens before corrosion, and m_1 is the average mass of the exposed part measured after each exposure time. Mass loss Δm can be expressed in net gram (g) or percentage (%). It can also be normalised by surface area and expressed in g/m^2 to eliminate the influence of differences in shapes and exposure areas of the test specimens.

From the mass loss, the dimension reduction, such as thickness loss, of the exposed part of the specimen for each exposure time can be determined as follows (Li et al. 2018):

$$C = \frac{\Delta m}{\rho_{st} A_s} \times 10^{-2} \tag{3.2}$$

where C is the thickness loss in millimetres (mm), Δm is the mass loss of the specimens in milligrams (mg), ρ_{st} is the steel density in g/cm^3, which is 7.85 g/cm^3 for mild steel, and A_s is the exposed area of the specimen in the acidic solution in cm^2 (i.e., the exposed part of the specimen). From the measured corrosion loss, the corrosion rate can be determined and expressed in various units as summarised in Table 2.2 of Chapter 2. In practice, the unit is usually mm/year.

Theoretically, the corrosion rate can also be determined from the mass loss, using Faraday's Law as follows (Wasim 2018):

$$C_r = \frac{K \cdot \Delta m}{A \cdot T \cdot D} \tag{3.3}$$

where C_r is the corrosion rate in mm/year, K is a constant equal to 8.76×10^4, Δm is the mass loss in grams, T is the exposure time in hours, D is the density in g/cm^3 of steel and A is the surface area of the specimen exposed in the acidic solution in cm^2.

Based on the measurement and calculation described above, the corrosion loss as a function of exposure time can be obtained. In the acidic solutions of HCl with three pH values, namely, pH = 5.0, 2.5 and 0.0, some results in terms of corrosion loss are shown in Figure 3.3, where each point is the average of three measurements of mass and thickness. It can be seen that the corrosion loss, both in mass loss and thickness loss, increases with immersion time almost linearly, in particular in acidic solutions with low acidity, i.e., pH = 2.5 and 5. The linear trend of corrosion loss is in line with some widely used models for corrosion loss, i.e., Equation (2.9) of Chapter 2. This trend, however, is different from the results of corrosion current, which is non-linear over time. The reason for this could be that the corrosion loss represents the cumulative effect of corrosion, which is more gradual, whilst the corrosion current represents

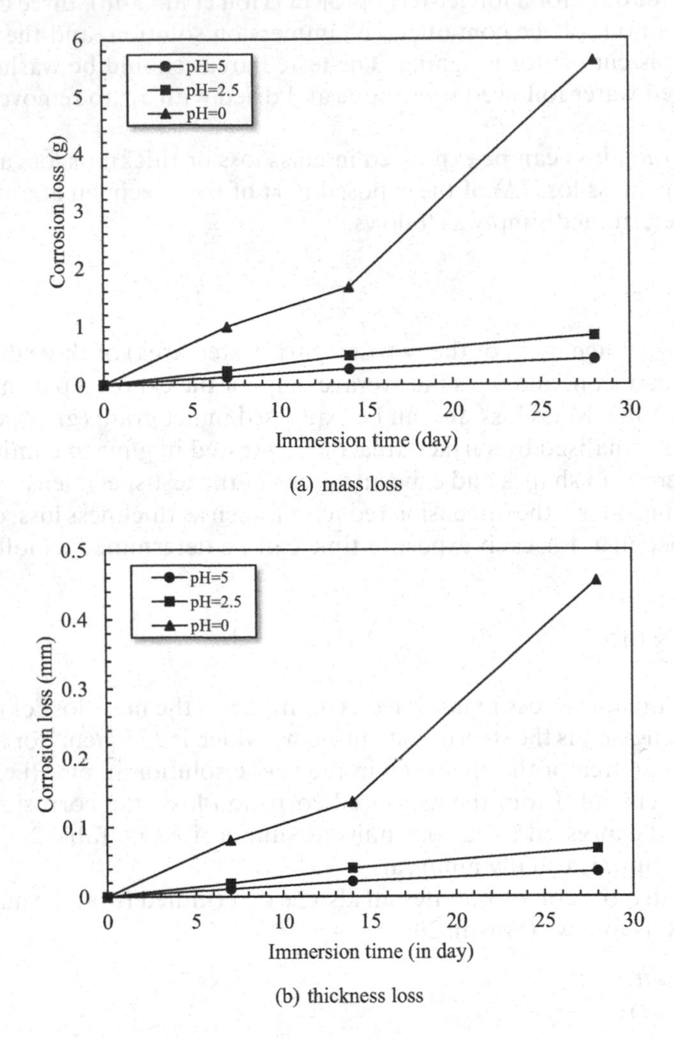

Figure 3.3 Corrosion loss from immersion tests: (a) mass loss and (b) thickness loss.

the instantaneous rate of corrosion, which is more fluctuated. It can also be seen that corrosion loss increases with the increase of acidity, i.e., reduction of pH. Clearly, a higher concentration of acid promotes the corrosion activities. The reason is that when the acid concentration increases, there are more hydrogen ion absorbed on the surface of steel, which in turn takes more electrons from the iron and hence accelerates corrosion. These results are consistent with most published results in the literature.

From Figure 3.3, it can be seen that corrosion loss as expressed in mass loss and thickness loss has the same trend. Based on this fact, expression of corrosion progress by mass loss or thickness loss can be exchangeable, although in practice, it is more expressed in thickness loss. Also, from Figure 3.3, the corrosion rate of the specimens can be determined which is the slope of the curves in Figure 3.3. Some results of corrosion rate of steel specimens in HCl solutions with various pH values are shown in

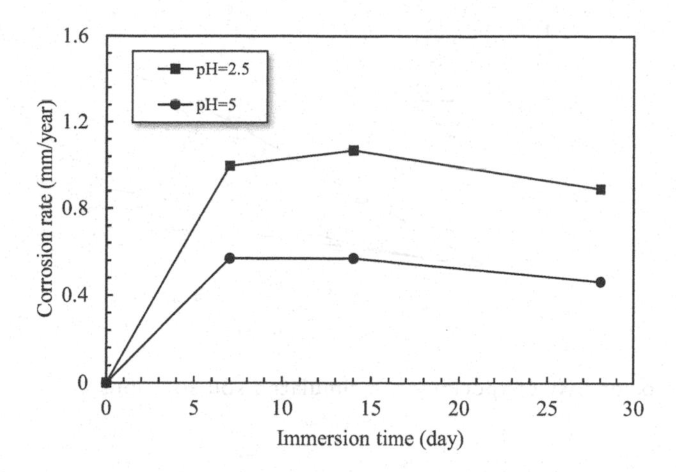

Figure 3.4 Corrosion rate of specimens in acidic solutions.

Figure 3.4. It can be seen from the figure that the corrosion rate increases sharply from the onset of corrosion, and after a whilst, it slows down and gradually decreases over time. This trend is very similar to the corrosion current in Figure 3.2. This is because corrosion current is the indication of instantaneous corrosion activities and hence the rate. The corrosion rate in Figure 3.4 is also similar to the conceptual model of corrosion rate presented in Figure 2.3 of Chapter 2, which consists of three stages: ascending, plateau and descending. There is a clear analogy in these two figures, providing some data as support of the concept expressed in Figure 2.3.

For the simulated soil solutions with various pH values, i.e., pH = 8.0, 5.5 and 3.0, the corrosion loss in terms of mass loss in g/m^2 is shown in Figure 3.5, where each point is the average of three measurements of mass loss. It can be seen that the mass loss increases with time almost linearly, which is similar to that in other acidic solutions as shown in Figure 3.3. It can also be seen that there is more mass loss or more severe corrosion in the solution with higher acidity, i.e., with a smaller pH value of 3.0. This is consistent with the results of corrosion current. As can be seen from the figure, there is not much difference in mass loss when the pH values are between 5.5 and 8.0, indicating that the effect of acidity starts to diminish once it is greater than, say, 5. As discussed in Section 2.3, once the pH value increases beyond 5, its effect on corrosion process starts to decrease. Results in Figure 3.5 provide some evidence for this analysis.

Whilst results from immersion tests can provide indication and perhaps trend of corrosion progress, as well as how environmental factors, such as pH, affect corrosion progress, for direct application of test results, the corrosion has to take place in a natural environment. As presented in Section 3.2.2, opportunity of three decommissioned steel structures and exhumed steel pipes is seized, and samples are extracted from these structures for analysis and testing. The corrosion loss over time for three comparable steel members is shown in Figure 3.6, where each point represents the average of more than three samples or measurements.

It can be seen from Figure 3.6 that corrosion loss of steel in a natural atmospheric environment is approximately linear in the same trend as that in corrosion immersion

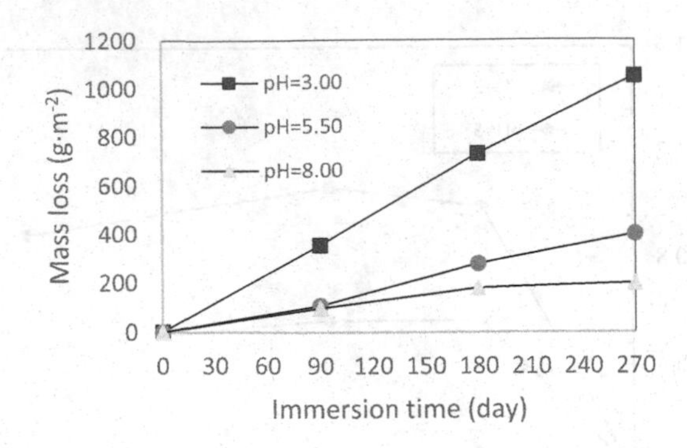

Figure 3.5 Corrosion loss of specimens in simulated soil solutions.

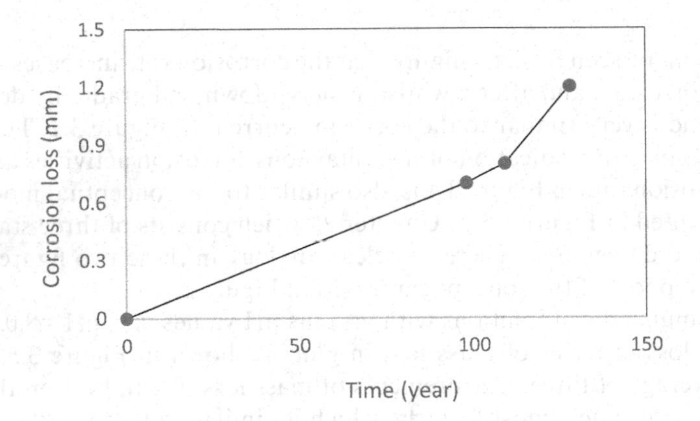

Figure 3.6 Corrosion loss of steel in natural atmosphere.

tests. These results can suggest that a linear relationship for corrosion loss over time can be used for steel structures exposed in a natural atmosphere. In fact, this is what many practitioners use in their design and assessment of steel structures exposed in a natural atmosphere (Li 2018). For the purpose of practical use or reference, an empirical formula can be derived from Figure 3.6 from regression analysis, as follows:

$$C = 7.3t - 3.6 \tag{3.4}$$

where C is the corrosion loss in μm, and t is natural time in years. The coefficient of determination for Equation (3.4) is 0.96. Obviously, it does not make sense that at time = 0, the corrosion loss is −3.6 μm, which should really be 0. Thus, the constant 3.6 in Equation (3.4) is for the purpose of best fitting the regression with a very high coefficient of determination (0.96), which makes sense after corrosion starts, i.e., $t > 0$. It can be seen that the corrosion rate, i.e., the slope in Figure 3.6, is 7.3 μm/year. This is well within the range of a large amount of data reported in the literature as presented in Chapter 2.

Table 3.3 Summary of 3D Scanning Results

Pipe Section No.	*Corrosion Pit Depth (mm)*		*Standard Deviation*
	Maximum	*Average*	
1.1	1.89	0.18	0.67
1.2	1.66	0.26	0.42
1.3	1.79	0.01	0.52
2.1	0.94	0.01	0.61
2.2	0.69	0.01	0.42
2.3	1.02	0.01	0.62
3.1	1.32	0.01	0.63
3.2	1.65	0.01	0.56
3.3	1.72	0.01	0.68

It may have been noted that the corrosion in immersion solutions and natural atmosphere is uniform as shown in Figures 3.3–3.6. This makes sense in that the environments that the specimens and steel are exposed to are quite "uniform". Since steel is also a quite unfirm material, the corrosion is uniform. In the real soil environment, however, the corrosion is non-uniform with more pitting. This again is understandable due to the fact that soil is complex both physically and chemically. Table 3.3 presents the corrosion pit depth for steel from the exhumed pipes which were obtained from 3D scanning, including the maximum pit depth and average pit depth. As expected, steel from exhumed aged pipes experienced various degrees of corrosion, which reflects the variations in material, exposure soil conditions and age. The maximum pit depth is 1.89 mm, but the pipe wall thickness is only 5 mm (see Table 3.2). The design service life of steel pipes can be 100 years. After 52 years, the corrosion pit penetrated nearly 40% of the wall thickness. Compared with corrosion in a natural atmosphere, corrosion in natural soil is more severe and more dangerous due to the pitting.

3.3 DEGRADATION OF TENSILE PROPERTIES OF STEEL

After steel is corroded, either from corrosion immersion tests or from natural corrosive environment as presented in Section 3.2, changes in the mechanical properties of the corroded steel can be studied. With the mechanistic approach adopted in this book, the concern is on the end results of corrosion after electrochemical reactions as measured by corrosion loss over time. Relative measurement of corrosion loss can provide some validity on examining the relative changes of the mechanical properties of the corroded steel. Such tensile properties are the most fundamental mechanical properties of steel in engineering practice for most industry, and they are examined first in this chapter.

Tensile tests are a very standard mechanical test for materials. It is relatively simple to perform, and the results of the test are very informative and useful. There are various standards to follow almost all of which are similar, meaning that the results are very similar no matter which standard to apply. ASTM E8/E8M (ASTM 2016) is one of the widely used standards and hence is followed in the tests presented in this section. A full-range stress and strain curve can be obtained for each specimen from tensile tests. Based on the stress–strain curve, the yield strength, ultimate strength and failure strain of the corroded steel can be determined.

Since the corrosion specimens are designed with consideration of subsequent tensile test, they can be used directly after a designated period of exposure. At the designated time, such as 7, 14 and 28 days for short-term tests (Li 2018) and 30, 90 and 270 days for longer-term tests (Hou et al. 2016) as described in Section 3.2.1, specimens are taken out of the container, i.e., the corrosive environment. Then corrosion measurements are taken and followed by tensile tests. Tensile tests are a very basic test for all materials, which will not be discussed here. Before the tensile testing, all dimensions of the tested part of the specimens (middle section of the dog-bone shaped specimen) should be measured for each specimen.

Conventionally, the stress and strain are determined based on original cross-sectional area of the specimen A_0 and gauge length L_0, denoted by σ_e and ε_e, respectively, from which a stress–strain curve is obtained, known as engineering stress–strain curve. The engineering stress σ_e and strain ε_e are mostly used in design and assessment of steel structures. In examining the corrosion effect on the mechanical properties of steel, however, it is more appropriate to use the true stress σ_t and strain ε_t (Garbatov et al. 2014) so that the corrosion-caused changes in cross section and subsequent length of the specimen in loading can be considered more accurately. The true stress σ_t and strain ε_t can be expressed as follows (Li 2018):

$$\sigma_t = \frac{P}{A_{acs}} = \frac{P}{A_0}(1+\varepsilon_e) = \sigma_e(1+\varepsilon_e) \tag{3.5}$$

$$\varepsilon_t = \int_{L_0}^{L} \frac{1}{L} \mathrm{d}L = \ln\frac{L}{L_0} = \ln(1+\varepsilon_e) \tag{3.6}$$

where P is the load corresponding to the displacement, A_{acs} is the actual cross-sectional area of the specimen and L is the actual gauge length. In Equations (3.5) and (3.6), $\sigma_e = \frac{P}{A_0}$ and $\varepsilon_e = \frac{\delta}{L_0}$, where δ is the displacement of the specimen measured in the tensile test. From Equations (3.5) and (3.6), a true stress–strain curve can be obtained. The yield strength is the stress at which a 0.2% offset line, i.e., plastic deformation at zero loading, intersects with the stress–strain curve. The ultimate strength is the maximum stress in tensile test, and failure strain is the strain when the specimen ruptures in the test. These are the basic mechanical properties of steel that are used in design and assessment of steel structures and hence in examining corrosion impact on their degradation in this chapter.

A typical true stress–strain curve for corroded steel from various corrosive environments is shown in Figure 3.7. It can be seen that no matter in what environment, the stress–strain curve of corroded steel is different from that of uncorroded steel. In general, the stress–strain curve of corroded steel is proportionally lower or smaller than that for uncorroded steel. This means that the tensile properties of the corroded steel have been degraded, including yield strength, ultimate strength, and failure strain, as to be analysed more quantitatively in the next sections. Results in Figure 3.7 are one of very few, if any, stress–strain curves for corroded steel in various environments. This is of practical importance with a view to design and assessment of steel structures in

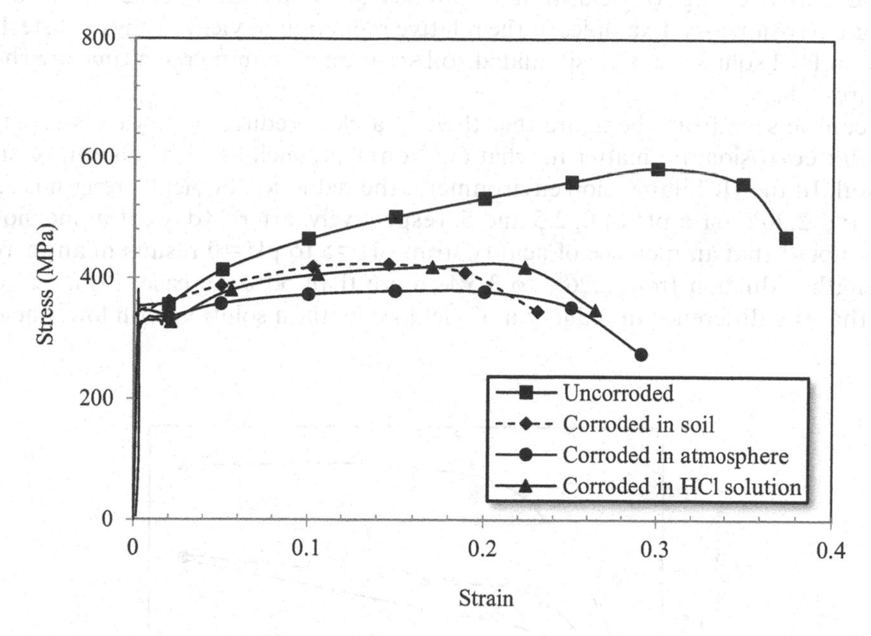

Figure 3.7 Stress–strain curve for corroded steel in various environments.

various corrosive environments. The significance of results in Figure 3.7 is more qualitative than quantitative indicating that different stress–strain curves should be used for steel in different corrosive environments.

3.3.1 Reduction of yield strength

Yield strength can be determined directly from tensile test or the true stress–strain curve for different corrosion conditions. The effect of corrosion on the reduction of yield strength can be expressed in a number of ways. The direct expression is actual reduction of yield strength with exposure time. It can also be expressed as the reduction of yield strength with corrosion loss. Also when test results are produced from the accelerated corrosion, such as immersion tests, the reduction of yield strength is better to be expressed in relative terms. This can eliminate the actual reducing process of the yield strength since only the resultant effect, i.e., end result, is compared with the original value. This relative approach is more appropriate for simulated corrosion and is adopted in the book as indicated in Section 3.2. In general, the relative change of a resultant effect X can be defined as follows:

$$\frac{x_0 - x_t}{x_0} \times 100 \tag{3.7}$$

where x_0 is the original value of the resultant effect or end result X, and x_t is the value of the end result X at time t. Obviously, when Equation (3.7) yields positive it is reduction and when it yields negative it is increase. Equation (3.7) will be used to calculate the relative change of corrosion effect in this book, such as reduction of yield strength.

The relative change of yield strength can be expressed with time, i.e., the exposure time, or corrosion loss. Examples of the relative reduction of yield strength of steel immersed in HCl solution and in simulated soil solution with immersion time are shown in Figures 3.8.

It can be seen from the figure that there is a clear reduction of yield strength of steel after corrosion, no matter in what environment, such as HCl solution or simulated soil. In the HCl immersion environment, the reduction of yield strength is 3.5%, 2.34% and 2.26% for a pH of 0, 2.5 and 5, respectively, after 28 days of immersion. It may be noted that an increase of acidity from pH = 5 to pH = 0 results in an increase of strength reduction from 2.26% to 3.5%, more than 50% increase. It may also be noted that the difference in reduction of yield strength in solution with lower acidity,

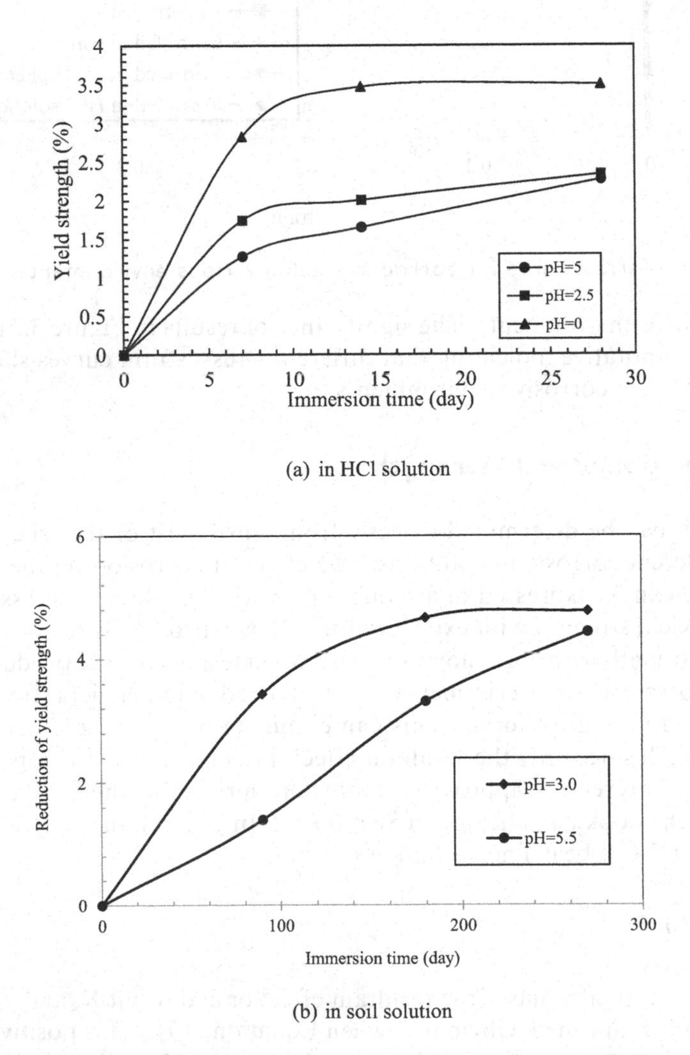

Figure 3.8 Reduction of yield strength with immersion time in (a) HCl solution and (b) simulated soil solution.

such as pH = 2.5 and 5, is small, such as less than 0.1%. This shows how aggressive the acidic solution is. The reason is again related to the hydrogen concentration. Lower pH means higher concentration of hydrogen and more hydrogen ions absorbed on the steel surface, which can take more electrons from the iron. As a result, corrosion is accelerated. It is of interest to note that after 14 days of immersion, the reduction rate of yield strength slows down, in particular for solution with pH = 0, which is almost flat (almost same reduction of 3.5%). For solutions with pH = 2.5 and 5, the reduction tends to be the same after 28 days of immersion. The reason for this can be that once steel is heavily corroded, such as in solution with pH = 0, the corrosion products cover the surface of steel, which prevents easy diffusion of oxygen to the surface and hence slows down the corrosion.

In the simulated soil environment, the reduction of yield strength is 3.44% and 1.39% for specimens in soil with a pH of 3 and 5.5, respectively, after 90 days, the difference of which is more than doubled. However, in the longer term, the effect of corrosion on yield strength of steel is almost the same. For example, at 270 days (9 months), the reduction of yield strength is almost the same in the soils with two quite different acidities, i.e., 4.77% reduction in the soil with pH = 3% and 4.43% in the soil with pH = 5.5. The reason can be that once steel is heavily corroded, the rusts cover the surface of steel. Thus, it is the diffusion of oxygen that controls the rate speed of corrosion and its effect on steel as well.

Reduction of yield strength versus mass loss in various exposure environments are shown in Figure 3.9: one in HCl solution and the other in simulated soil solution. A similar trend of yield strength reduction can be observed to that of Figure 3.8. To be specific, after 28 days exposure, yield strength decreases from 342.57 to 334.56 MPa in HCl solution with pH = 5 when mass loss reaches 1.33%, to 333.58 MPa in HCl solution with pH = 2.5 when mass loss reaches 2.55% and to 332.79 MPa in HCl solution with pH = 0 when mass loss increases to 17.7%. In Figure 3.9b, the reduction of yield strength is expressed in percentage and the mass loss in actual g/m^2. As can be seen, in a more acidic soil solution with pH = 3, a mass loss of 200 g/m^2 results in a reduction of 5.4% in yield strength of the steel, whilst in the less acidic soil soliton (pH = 5.5), about 400 g/m^2 of mass loss is needed to induce a similar reduction of yield strength, i.e., 4%. The results suggest that it is not just the corrosion loss (mass) that can determine the degradation of yield strength. The corrosion process which may alter the microstructure of the steel may have some impact as well.

The thickness loss of steel due to corrosion is more frequently used in practice as an indicator of corrosion risk. Figure 3.10 shows the relative reduction of yield strength in % versus thickness loss in both immersion environment and natural environment. The reduction of yield strength due to natural corrosion is obtained from samples extracted from the decommissioned steel structures as presented in Section 3.2.2. As can be seen in Figure 3.10, at the corrosion loss of 1.4 mm, the reduction of yield strength in HCl immersion is 6.2%. It is of interest to see that, at the corrosion loss of 3 mm, the reduction of yield strength in the natural environment is 6.1%, which is about the same as that in HCl immersion at the corrosion loss of 1.4 mm (but about half of the corrosion loss). This result again suggests that accelerated corrosion can also accelerate the effect on the mechanical properties of corroded steel. Evidently, the latter has not been accorded much attention. Thus, results in Figure 3.10, and in fact in other figures presented in the chapter are not widely available but of great practical significance.

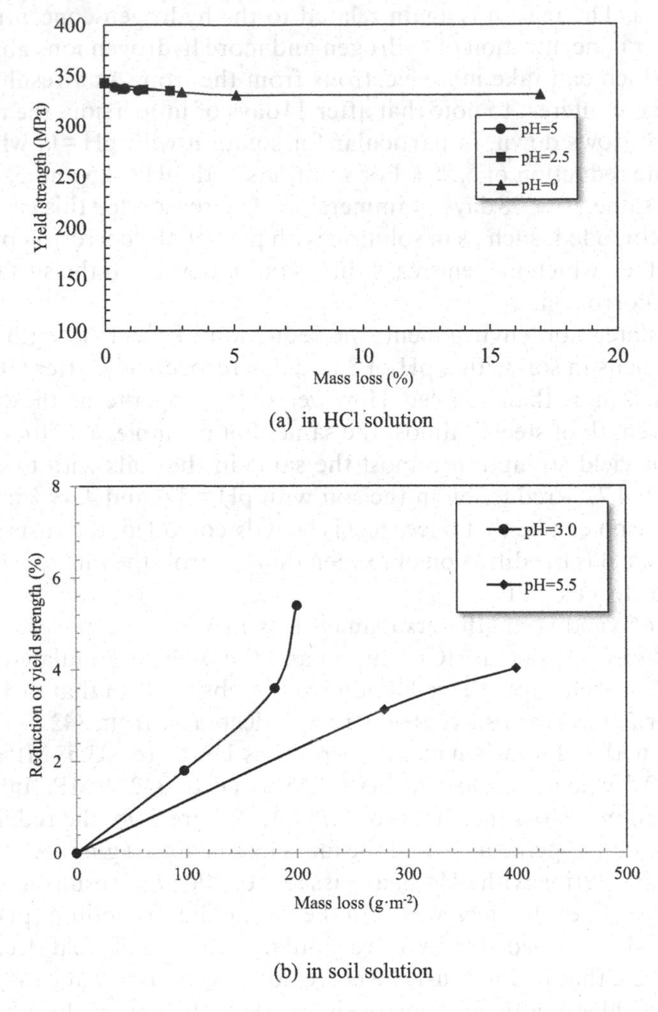

Figure 3.9 **Reduction of yield strength with mass loss in (a) HCl solutions and (b) simulated soil solution.**

Since there are two methods to calculate the stress and strain: engineering stress and strain and true stress and strain as shown in Equations (3.5) and (3.6), it would be of interest and practical significance to see the difference in corrosion effect on mechanical strength between the two methods of calculation. Figure 3.11 shows the reduction of yield strength calculated by the two methods for steel corroded in different environments. It can be seen that the level of reduction is higher for true yield strength than engineering yield strength of corroded steel in both environments. The difference can be as high as 20%, which is considerable. This phenomenon is also observed for corrosion in natural environment as shown in Figure 3.11b. It can be seen that the level of reduction is higher for true yield strength than engineering yield strength at the same exposure time. The difference is about 10%, which can be very significant for

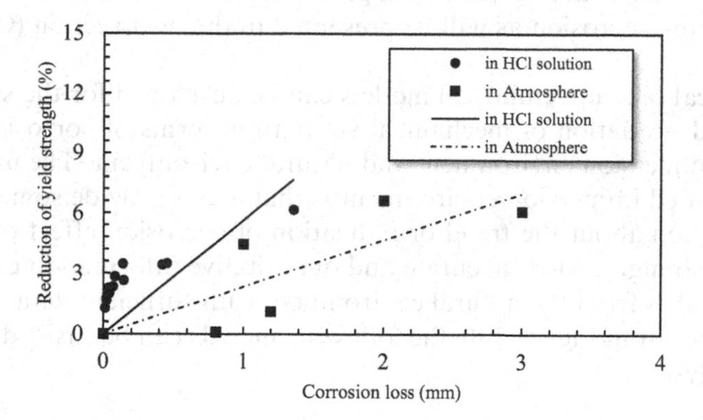

Figure 3.10 Reduction of yield strength with thickness loss in different environments.

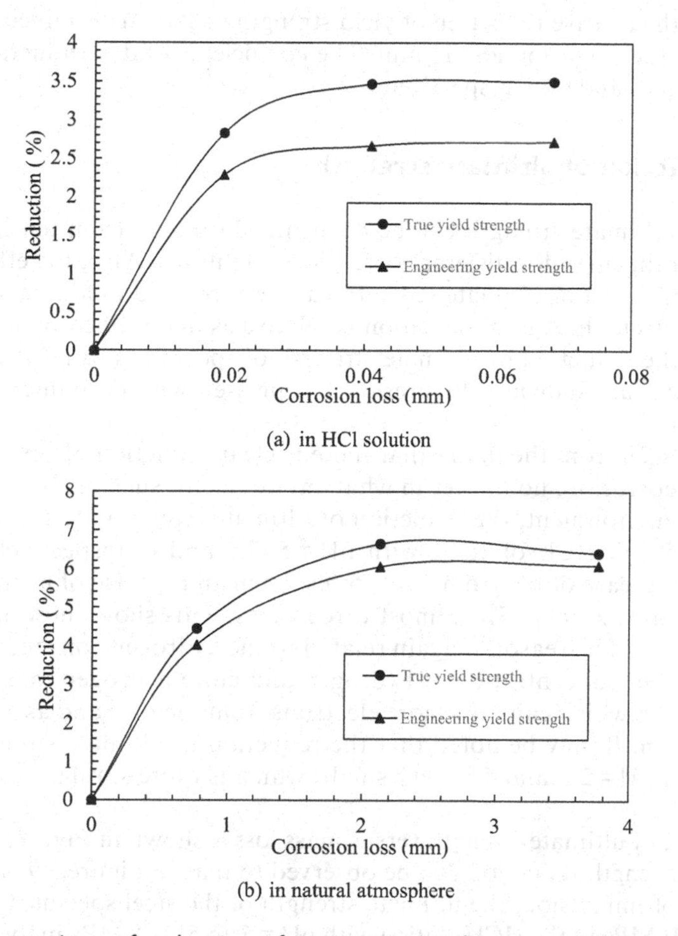

Figure 3.11 Comparison of reduction of engineering yield stress and true yield strength in (a) HCl solution and (b) natural atmosphere.

assessment of corroded steel structures in practice. The reason could be due to the loss of ductility during corrosion as will be presented in the next section (Garbatov et al. 2014).

For practical take up, empirical models can be developed for the study of corrosion effect on degradation of mechanical strength in terms of corrosion loss in mm for simulated immersion environment and natural environment. The usefulness for a model in simulated immersion environment is that it can provide some kind of qualitative information about the trend or indication of corrosion effect on degradation of mechanical strength. More accurate and quantitative information can only be obtained from models from the natural environment with sufficient data which may not be unique either. From Figure 3.10, the following models can be easily developed from regression analysis:

$$\Delta\sigma_y = 5.64C \text{ for corrosion in acidic solution} \tag{3.8a}$$

$$\Delta\sigma_y = 2.33C \text{ for corrosion in atmosphere} \tag{3.8b}$$

where $\Delta\sigma_y$ is the relative reduction of yield strength in % as determined by Equation (3.7), and C is the corrosion loss in mm. The coefficients of determination for Equations (3.8) are 0.66 and 0.85, respectively.

3.3.2 Reduction of ultimate strength

Likewise, the ultimate strength can be determined directly from tensile test or the true stress–strain curve for different corrosion conditions. Also, the effect of corrosion on the reduction of ultimate strength can be expressed in a number of ways as that for yield strength. Again, reduction is relative as determined by Equation (3.7). Examples of the reduction of ultimate strength of specimens in HCl solutions with various acidities are shown in Figures 3.12, expressed with (a) immersion time and (b) mass loss.

It can be seen from the figure that there is clear reduction of ultimate strength of steel after corrosion, no matter in what environment, such as different acidities. In the acidic environment, the reduction of ultimate strength after 28 days is 4.2%, 4.7% and 13.5% in HCl solutions with pH = 5, 2.5 and 0, respectively. It may be noted that a decrease of pH from 5 to 2.5 results in an increase of ultimate strength reduction from 4.7% to 13.5%, almost three times. This shows how aggressive the acidic solution is. The reason is again related to the hydrogen concentration. Lower pH means higher concentration of hydrogen and more hydrogen ions absorbed on the steel surface, which can take more electrons from the iron and as a result, accelerates corrosion. It may be noted that the reduction in ultimate strength in acidic solutions with pH = 2.5 and 5 is very small, which is quite similar to that for yield strength.

Reduction of ultimate strength versus mass loss is shown in Figures 3.12b. A similar trend of strength reduction can be observed to that in Figure 3.9. To be specific, after 28 days of immersion, the ultimate strength of the steel specimen reduces from 537.75 to 515.31 MPa in the HCl solution with pH = 5, to 512.58 MPa in the HCl solution with pH = 2.5 and to 465.16MPa in the HCl solution with pH = 0. It may be of interest

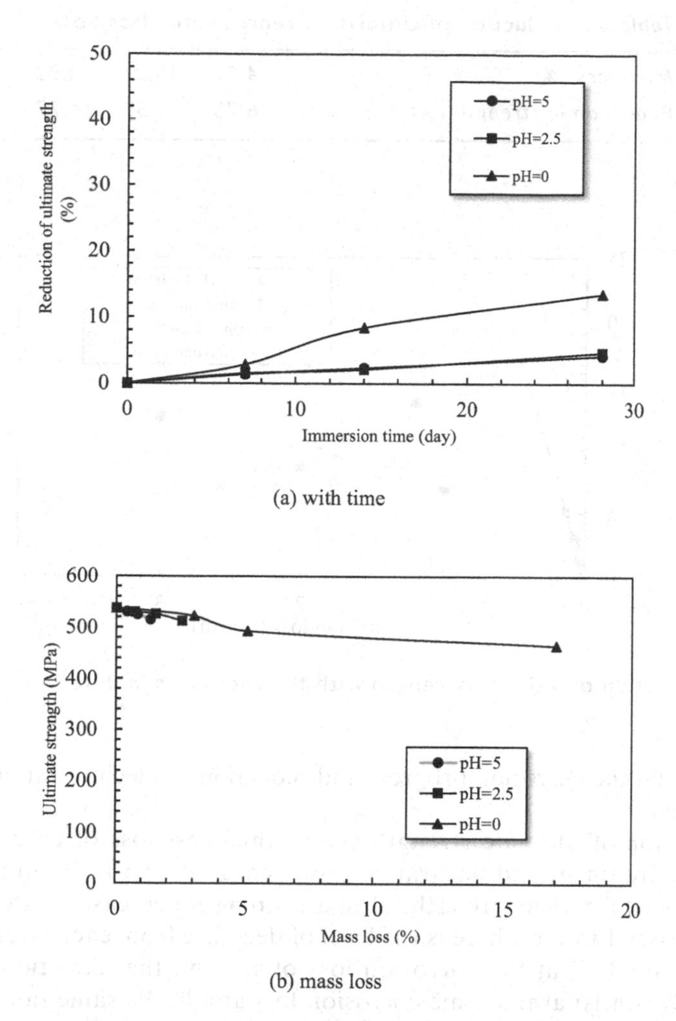

Figure 3.12 **Reduction of ultimate strength in HCl solutions with (a) immersion time and (b) mass loss.**

to see that in the same HCl solution, the reduction of yield strength is smaller than that of ultimate strength. For example, in the HCl solution with pH = 5, the reduction of ultimate strength is from 537.75 to 515.31 MPa, i.e., about 4.4%, after 28 days of immersion, whilst the reduction of yield strength is about 2.4% as shown in Figure 3.8. The difference is almost double. The reason for this difference could be that larger stress leads to more corrosion, given other conditions the same. The effect of applied stress on corrosion will be discussed in Chapter 5 in detail.

The reduction of ultimate strength versus mass loss in the natural atmosphere is presented in Table 3.4. It can be seen that the reduction of ultimate strength with mass loss is faster initially than later. For example, when mass loss is 4.81%, the reduction of ultimate strength is 6.75%, but when mass loss is 28.92%, which is more than five times higher, the reduction of ultimate strength is 14.77%, which is only two times higher.

Table 3.4 Reduction of Ultimate Strength with Mass Loss

Mass loss (%)	4.81	15.21	28.92
Reduction in strength (%)	6.75	8.59	14.77

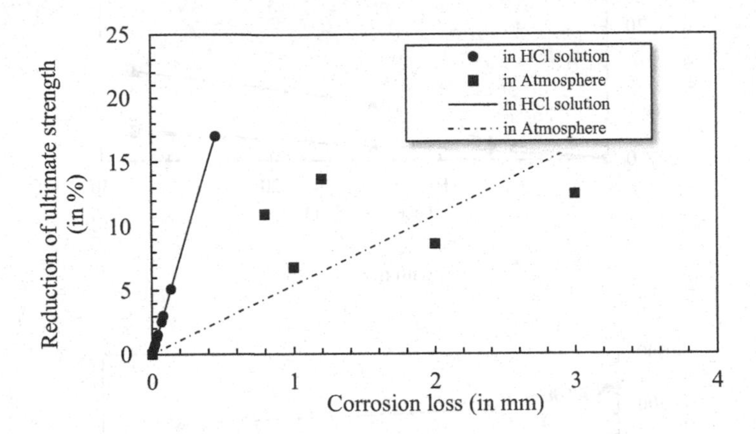

Figure 3.13 Reduction of ultimate strength with thickness loss in different environments.

This suggests that the corrosion progress and reduction of mechanical properties are not proportional.

The reduction of ultimate strength versus thickness loss of corroded steel in both acidic environment and natural environment is shown in Figure 3.13. In the same trend as that for yield strength, ultimate strength decreases with the increase of thickness loss. Of interest here is the level of decrease from each strength. As can be seen in Figure 3.13, at the corrosion loss of 1.4 mm, the reduction of ultimate strength is 28%, whilst at the same corrosion loss and in the same immersion solution, the reduction of yield strength is 6.2% (Figure 3.10). This is about 4.8 times more. It can be seen from Figure 3.13 that at the corrosion loss of 3 mm, the reduction of ultimate strength in the natural environment is about 15%, which is more than twice that of yield strength at the same corrosion loss (Figure 3.10). Compared with the reduction rate of ultimate strength in HCl solution, again the ultimate strength degrades at a slower rate in natural environment than that in accelerated environment, as can be seen from their slopes. Again, this phenomenon has not been widely observed or reported. Obviously, using results from simulated immersion tests in practical design and assessment of steel structures requires extreme caution not just for corrosion loss but also for degradation of mechanical strength of the corroded steel.

The difference in corrosion effect on mechanical strength between two methods of calculations, i.e., true strength and engineering strength, is also observed for ultimate strength as that for yield strength, but the comparison is omitted here since they are similar or same as that of yield strength as least in trend. It is therefore imperative that true mechanical properties are used in prediction of corrosion effect on the mechanical properties of steel.

Following the approach for developing models for yield strength reduction, models for ultimate strength reduction can also be developed from regression analysis of the data in Figure 3.13 as follows:

$$\Delta\sigma_u = 37.33C \quad \text{for corrosion in acidic solution} \tag{3.9a}$$

$$\Delta\sigma_u = 5.38C \quad \text{for corrosion in atmosphere} \tag{3.9b}$$

where $\Delta\sigma_u$ is the relative reduction of ultimate strength in % as determined by Equation (3.7), and C is the corrosion loss in mm. The coefficients of determination for Equations (3.9) are 1.0 and 0.8, respectively.

Figure 3.14 shows the reduction of ultimate strength of the corroded steel, cut from a pipe exhumed from service in natural soil. To make the results comparable amongst pipes with different thicknesses, the corrosion depth, or thickness loss a is divided by the pipe wall thickness t, i.e., $a/t \times 100\%$. Both the average corrosion depth and the maximum corrosion depth (pit) are considered. It can be seen that the reduction of ultimate strength is different for pitting corrosion and uniform corrosion as expressed in maximum corrosion depth and average depth. In general, a relationship between the reduction of ultimate strength of corroded steel and corrosion loss can be observed with a relatively strong R^2 value (i.e. 0.76) for maximum corrosion pit depth, whilst R^2 for the average corrosion loss is only 0.37. This implies that the reduction of ultimate strength is more sensitive to deeper corrosion pit than the average pit depth of corrosion.

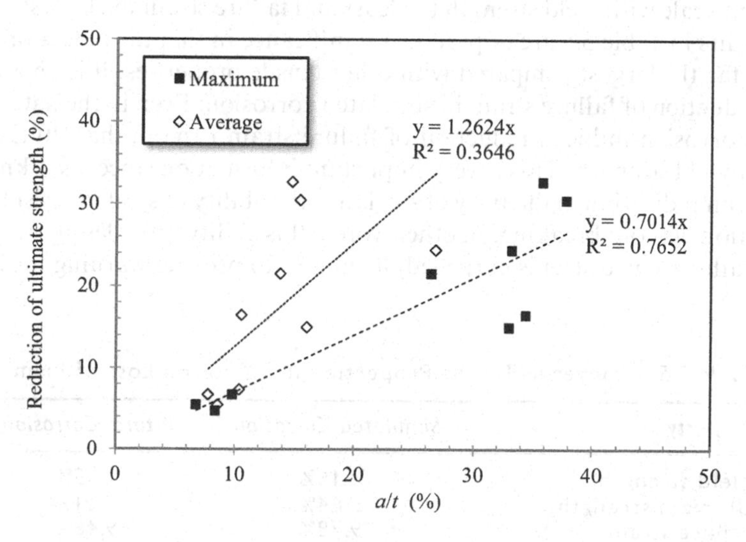

Figure 3.14 Reduction of ultimate strength with thickness loss in natural soil.

3.3.3 Reduction of failure strain

The failure strain is the strain at which the steel ruptures in the tensile test. It is an indicator of the ductility of steel. When steel suffers from corrosion, its failure strain reduces as well in a similar manner to its strength. From the stress–strain curve (such as Figure 3.7), the failure strain of the steel exposed in different environments and for different periods of exposure time can be obtained. The reduction of failure strain can be expressed in terms of exposure time or corrosion loss as those for strength. Figure 3.15 shows the reduction of failure strain of specimens immersed in different environments, expressed in terms of immersion time in days and corrosion loss in mm. From the figure, a similar trend to those for yield and ultimate strength can be observed. For example, in general, reduction of failure strain increases with the increase of corrosion no matter how it is expressed. Also the higher the acidity of the solution, the more the reduction in failure strain as shown in Figure 3.15a and b.

Also, the reduction of failure strain for steel corroded in natural atmosphere extracted from the decommissioned steel structures is shown in Figure 3.15c. As can be seen, with the increase of corrosion in terms of thickness loss, the reduction of failure strain increases. At the thickness loss of 1, 2 and 3 mm, the reduction of failure strain is about 15%, 22% and 29%, respectively. This indicates that the reduction is relatively more for less corrosion loss.

The changes in yield strength, ultimate strength and failure strain at 1 mm of corrosion loss are compared between the simulated corrosion and natural corrosion, as shown in Table 3.5. It can be seen that at the same degree of corrosion, the level of reduction of tensile properties is generally higher for steel subjected to simulated corrosion than natural corrosion. This is because there is more hydrogen absorbed into the steel by immersing steel in HCl solutions. Therefore, hydrogen embrittlement is more severe by immersing steel in HCl solution. In addition, chloride atoms in the solution can break the passive film and lead to the formation of pits and cracks during corrosion (Revie and Uhlig 2008). Also can be seen is that reduction of tensile properties caused by corrosion is of different scale with yield strength the least and failure strain the largest.

The results in Table 3.5 are of practical significance in that the reduction of failure strain is by far the largest compared with other tensile properties. It is about 20 times larger for reduction of failure strain in simulated corrosion. Even in the natural atmosphere, the corrosion-induced reduction of failure strain is more than 10 times larger than that of yield strength. This is very important information since, as is known, failure strain is an indication of ductility of steel. It is the ability of steel to resist the stretch or deformation without breaking. In other words, this ability provides pre-warnings for structural failure. Once steel is corroded, its ability to provide warning for its failure,

Table 3.5 Changes in Tensile Properties at Corrosion Loss of 1 mm

Property	*Simulated Corrosion*	*Natural Corrosion*
Yield strength	4.15%	1.55%
Ultimate strength	25.64%	9.01%
Failure strain	79.78%	19.48%

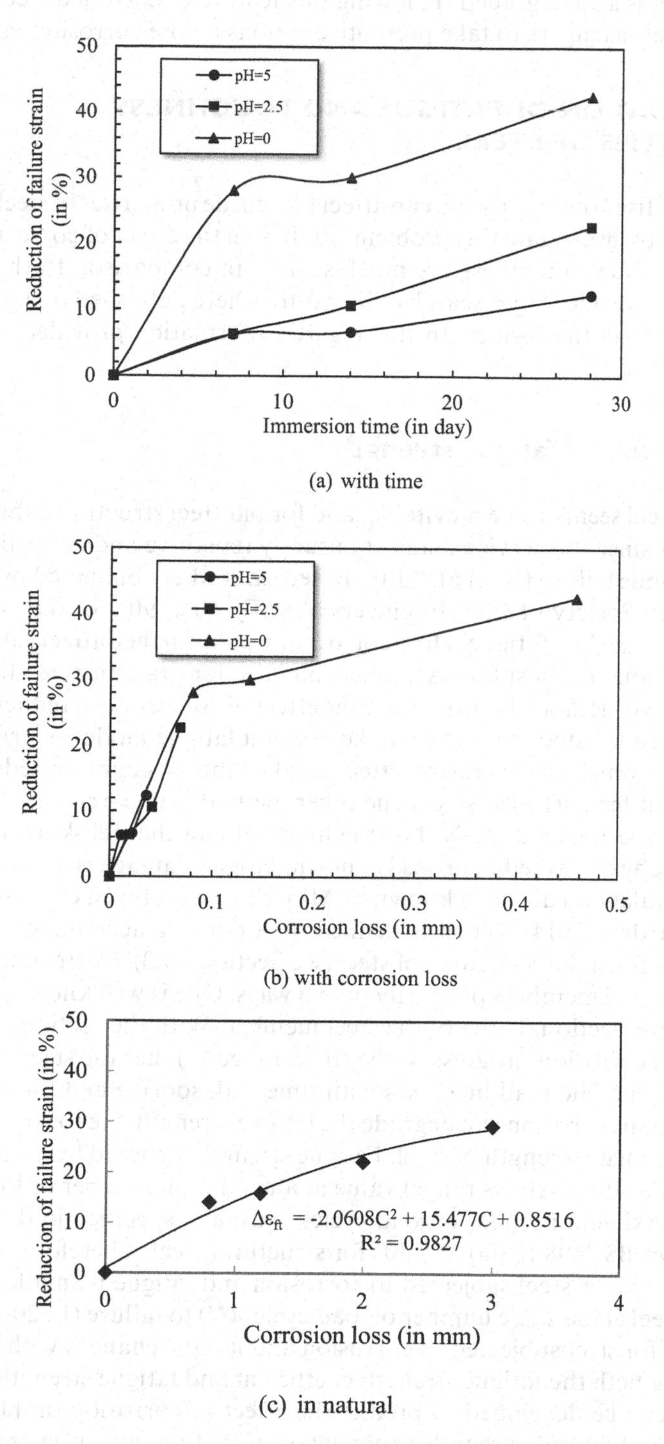

Figure 3.15 Reduction of failure strain in HCl solution with (a) time, (b) thickness loss and (c) in natural atmosphere.

such as rupture, is also degraded. Knowing this feature of corroded steel can help engineers and asset managers to take preventive actions before corrosion takes its tolls.

3.4 DEGRADATION OF FATIGUE AND TOUGHNESS PROPERTIES OF STEEL

If it has been realised that corrosion can affect the tensile properties of steel and research is being started to investigate this problem, studies on the effect of corrosion on fatigue strength and fracture toughness are much scarcer in comparison to those on tensile properties. This is evident by a search of literature where published papers on the latter are much more than the former. In this regard, information provided in this section can be quite useful.

3.4.1 Reduction of fatigue strength

Corrosion of steel seems to be inevitable, and for old steel structures, the problem can be exacerbated since the service loads are usually repetitive and/or cyclic and eventually lead to fatigue failure (Ni et al. 2010). Based on studies conducted by the Committee of American Society of Civil Engineers (ASCE), 80%–90% of the failures in steel structures are related to fatigue. How corrosion would further affect fatigue failure is of great significance not just for researchers but also for practitioners alike.

There are two methods to investigate the effect of corrosion on reduction of fatigue strength of corroded steel. One is to make use of a fatigue model as briefly discussed in Chapter 2, in which the corrosion effect is taken into account on reduction of ultimate strength of the corroded steel. The other method is to carry out fatigue tests on corroded steel specimens directly. For the first method, the well-known Miner's rule is used and briefly described here. As is known, fatigue damage is assessed by a linear damage accumulation rule, also known as Miner's rule (Nguyen et al. 2013, Adasooriya and Siriwardane 2014), which is defined by a damage accumulation index to be determined based on the S-N curve of steel (see Section 2.1.3). Corrosion affects the fatigue failure of steel members primarily in two ways. One is well known, i.e., corrosion reduces the cross-sectional area of the steel member. With the sectional area reduced over time when corrosion progresses, the stress range (S) that the steel member is subjected to under a cyclic load increases with time (Adasooriya and Siriwardane 2014). The second is that corrosion can degrade the fatigue strength due to corrosion-induced reduction of ultimate strength of steel. Fatigue strength, denoted by σ_f in this chapter, can be defined as the S (stress range) value at a maximum number of load cycles N at failure. The maximum number of load cycles is usually specified in design standard. For example, in BS 7608 (2014), $N = 10^7$ for structural steel. Therefore, the magnitude of stress range (S) for steel subjected to corrosion and fatigue is smaller than that for uncorroded steel at the same number of load cycles (N) to failure (Li 2018). As a result, the S-N curve for steel subjected to corrosion and fatigue changes with corrosion and time, including both the fatigue strength coefficient and fatigue strength exponent.

A model can be developed to predict the effect of corrosion on fatigue strength coefficient A and fatigue strength exponent B, as a function of corrosion rate and time. Field studies (Li 2018) show that most fatigue occurs under cyclic normal stress.

Therefore, this section only covers the corrosion effect on the *S-N* curve of steel subjected to normal stress. As discussed in Chapter 2, fatigue under cyclic shear stress is quite a different phenomenon, which will not be covered in this book.

As presented in Section 2.1.3 briefly, the fatigue capacity of steel is represented by the *S-N* curve, which is largely based on experiments and can be expressed as follows (Zhao 1995):

$$N = A\,S^{-B} \tag{3.10}$$

where N is the number of load cycles to failure, S is the stress range, A is the fatigue strength coefficient and B is the fatigue strength exponent. Both A and B are related to the tensile properties of steel and can be expressed as follows (British Standards Institution 2014, Zhao 1995):

$$A = N_1(k_r\sigma_u)^B \tag{3.11}$$

$$B = \frac{\log N_0 - N_1}{\log(k_r\sigma_{u0})} \tag{3.12}$$

where N_0 is the largest load cycle in the low load cycle region, N_1 is the minimum load cycle in the high load cycle region, σ_u is the ultimate strength of steel, k_r is the ratio of fatigue strength limitation to ultimate strength and σ_{u0} is the original ultimate strength of steel (Zhao 1995, Bandara et al. 2015). A log plot of Equation (3.10) is the well-known *S-N* curve subjected to normal stress (S).

In Equations (3.11) and (3.12), parameters N_0, N_1 and k_r are specified in codes and standards. For the example of BS 7608 (British Standards Institution 2014), $N_0 = 10^5$, $N_1 = 10^7$ and $k_r = 0.31$. Due to the corrosion of steel, the ultimate strength σ_u changes with corrosion over time. As such, the *S-N* curve of corroded steel will change with corrosion over time. If a relationship between ultimate strength of corroded steel and corrosion loss can be developed, A and B can be determined, and the *S-N* curve can be established. For example, from Equation (3.9b), the ultimate strength after corrosion at time t can be expressed as follows:

$$\sigma_u(C) = \sigma_{u0}(1 - 0.054C) \tag{3.13}$$

where $\sigma_u(C)$ is the ultimate strength of corroded steel in MPa after corrosion loss of C in mm in natural atmosphere. With Equation (3.13), the corrosion effect on fatigue can be determined. An example of the *S-N* curve for G250 steel (σ_{u0}= 430 MPa) with different corrosion losses such as C=0, 3, 6 mm is shown in Figure 3.16, where $N_0 = 10^5$, $N_1 = 10^7$ and $k_r = 0.31$, based on BS 7608 (British Standards Institution 2014).

Since the fatigue strength σ_f is defined as the stress range S value at $N = 10^7$ (BS 7608 2014), the reduction of fatigue strength can be obtained directly from Equation (3.13) with given k_r as follows:

$$\sigma_f = k_r\sigma_u(C) \tag{3.14}$$

The fatigue strength of steel with various corrosion losses as determined by Equation (3.14) is shown in Figure 3.16 on the right-hand side of the figure.

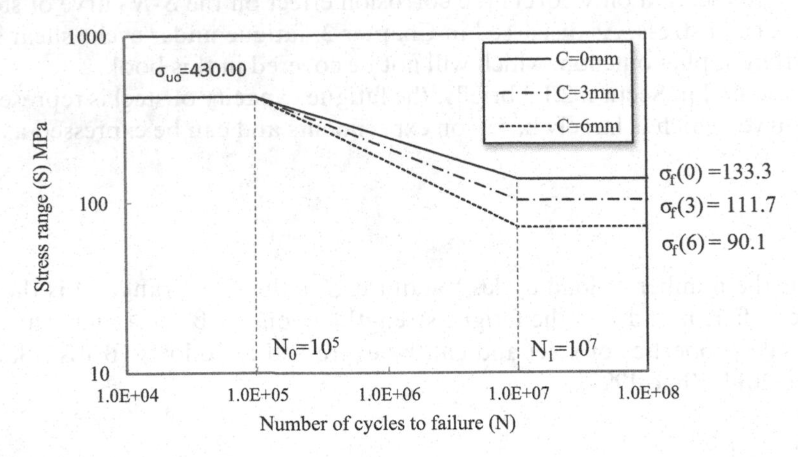

Figure 3.16 S-N curve for steel with various corrosion losses.

The second method is to carry out fatigue tests directly on corroded steel. The *S-N* curve of corroded steel can be determined from constant amplitude axial fatigue tests on corroded steel. The design of the fatigue test should follow a standard, such as ASTM E466-15 (ASTM 2015). There is a difference in considering the corrosion effect on fatigue from that on other mechanical properties, such as tensile strength whereby steel is corroded first, and the mechanical tests are carried out subsequently. For corrosion effect on fatigue, this may not be appropriate since fatigue is a long-term effect, and during the process of fatigue, corrosion would interact with fatigue. Ideally, all corrosion effects on mechanical properties should be considered simultaneously with the corrosion process because there may be more or less interaction between corrosion and applied loads and even the changes of mechanical properties. Whilst it may be tolerable or even acceptable that simultaneous corrosion and mechanical tests may not be essential for some mechanical properties, such as tensile strength, the effect of corrosion on fatigue strength needs to be considered in conjunction with the fatigue process to be more accurate. For this reason, only test results on fatigue from corroded steel taken from the decommissioned steel structure is presented since the steel was subjected to simultaneous corrosion and cyclic loading, i.e., fatigue.

For accurate tests on corrosion effect on fatigue, specimens are best taken from a decommissioned steel structure after certain years of service under simultaneous corrosion and cyclic loading. As described in Section 3.2.2, corroded steel members need to be classified in terms of corrosion loss in thickness. To study the effect of corrosion, samples should be taken from each level of corrosion loss, and due to variations in sampling and testing, a number of samples, such as 10 samples, should be prepared for fatigue tests. In addition, before fatigue tests, indeed all mechanical tests are carried out, and all corroded samples should be rinsed in rust removal solution and then ground with superfine sandpaper (600 grit) to remove the rust in accordance with a standard, such as ASTM G1-03 (ASTM 2017d). Samples are then cleaned thoroughly with bi-distilled water, followed by acetone and dried with air. Both ends of the samples need to be polished to fit the clamp systems of the fatigue test machine. The prepared samples are then placed under axial constant cyclic load with a frequency of 30 Hz with a number of stress ranges following ASTM E466-15 (2015).

For the example of decommissioned steel structures described in Section 3.2.2, the samples are taken from corroded steel members with the thickness of 15 mm for mildly corroded steel plates, 12 mm for moderately corroded plates and 10 mm for severely corroded plates. The configuration of the test samples is determined based on their thickness as per ASTM E466-15 (2015).

According to the tensile test results (Section 3.3.2), the original ultimate strength of the steel (σ_{u0}) is 420 MPa. Therefore, the stress ranges exerted on the samples are 0–100, 0–120, 0–220, 0–320 and 0–420 MPa (ASTM E466-15 2015). Three duplicate samples are tested for each stress range, and the number of cycles to failure for each sample is recorded. The stress range *S* and the number of cycles to failure *N* for each sample can be plotted. A plan for the fatigue can be developed as shown in Table 3.6.

Results of the *S-N* curve for different corrosion losses are shown in Figure 3.17 (Li et al. 2019c). It can be seen that with the increase of corrosion loss, the *S-N* curve shifts downwards with the decrease of both stress range (*S*, such as fatigue strength) and number of load cycles. This clearly suggests that the fatigue resistance of corroded steel is reduced. It can be seen that at the small level of corrosion, such as $C<3$, the

Table 3.6 Design of Fatigue Test

Specimen Condition	*Stress Range (MPa)*
Mild corrosion	0–160
C = 0.77 mm	0–180
Moderate corrosion	0–220
C = 2.13 mm	0–320
Severe corrosion	0–420
C = 3.76 mm	

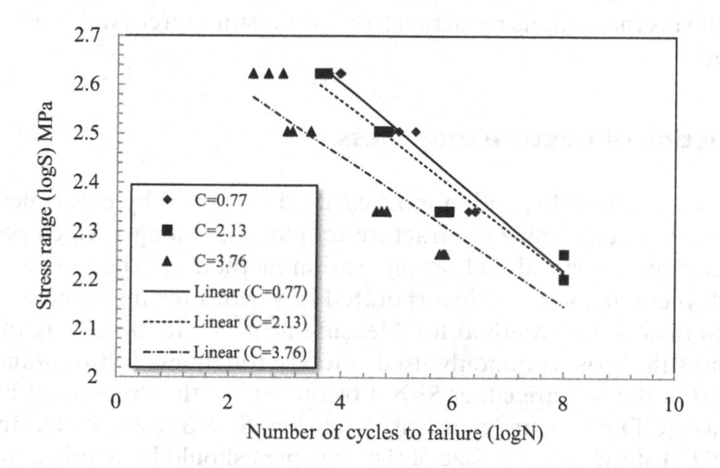

Figure 3.17 Changes in S-N curves with corrosion progress.

Table 3.7 Maximum Stress and Number of Load Cycles at Failure

Corrosion Loss (mm)	*Stress at Failure (MPa)*	*Number of Load Cycles at Failure*
C = 0.77	218.78	2,511,886
C = 2.13	182.39	1E+08
C = 3.76	181.97	501,187

difference in reduction of fatigue strength is small. At $C=3.76$ mm, the fatigue strength reduces from about 263.03 to 199.53 MPa at the load cycle of 10^6, which is a very significant reduction and will surely severely affect the safety and serviceability of corroded steel and steel structures. Furthermore, for the same level of fatigue strength (i.e. S), the number of load cycles reduces from 10^8 to $10^{5.7}$ (501,187 cycles) when the corrosion loss increases to 3.76 mm. Again, this is very significant as far as structural safety is concerned.

Since the fatigue strength (σ_f) is defined as the maximum stress at the number of load cycles at failure, it can be obtained from the test results of Figure 3.17. Table 3.7 presents the stress at fatigue failure and the corresponding number of load cycles for various corrosion losses. It can be seen that for an increase of corrosion loss from 0.77 to 3.76, the stress range at failure, i.e., the fatigue strength reduces from 218.78 to 181.97, which is 16.83% reduction. In addition, the number of load cycles reduces from 2511,886 to 501,187, which is 80% reduction. This suggests that the fatigue is the interaction of stress and number of load cycles. They do not change in the same proportion. The corrosion effect on them is not the same either.

It needs to be noted that both methods to predict reduction of fatigue strength would yield similar results if not the same. For example, Figure 3.18 shows the comparison of the S-N curve determined from the prediction model based on corrosion loss (Equation 3.14) with that directly from the fatigue test results with moderate corrosion (Li et al. 2019). The R^2 value (coefficient of determination) between the predicted S-N curve and the test results is 0.96 for median corrosion, and the p value (significance) is 1×10^{-8}. This means that fatigue reduction from test results agrees well with that of the prediction model.

3.4.2 Reduction of fracture toughness

Fracture toughness is usually, and in most cases, determined by experiments. There are many methods to determine the fracture toughness, amongst which perhaps the unloading compliance method and normalisation method are the most widely used methods. Both these methods are incorporated in various testing standards, such as ASTM 1820 Standard Test Method for Measurement of Fracture Toughness (2014) which is perhaps the most commonly used fracture toughness testing standard. The specimens used in the test are either SENB beams under the three-point bending or compact tension (CT) plates under eccentric tension. SENB test is much simpler that CT test. In ASTM standards, the size of the specimen should be within a limit with a single edge notch at the centred and should be fatigue-cracked in three-point bending

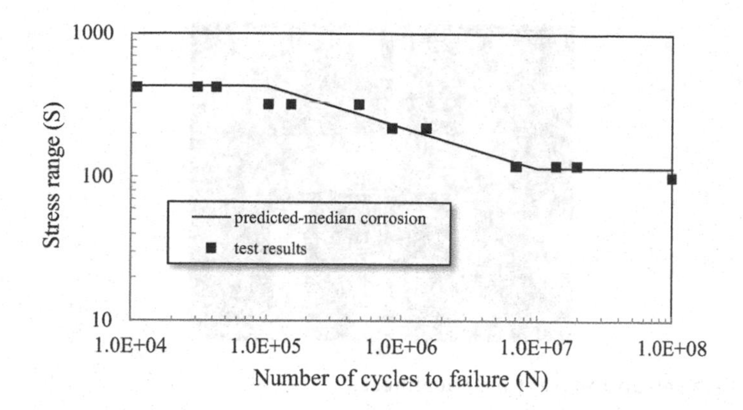

Figure 3.18 Comparison of S-N curves from prediction model with test results.

with a support span, S, nominally equal to four times the width, W. A key in specimen design is to determine its width to ensure plane strain condition with the formula as follows (ASTM E1820 2018b):

$$B \geq 2.5\left(\frac{K_{IC}}{\sigma_y}\right)^2 \tag{3.15}$$

where B is the width of the specimen, K_{IC} is the fracture toughness and σ_y is the yield strength of the material. When Equation (3.15) is not met, the specimen will facture in the plane stress condition. For the plane stress fracture, the fracture toughness decreases with the increase of specimen width and stabilises at a certain width at which plane strain fracture occurs. However, for structural steel, the calculated width for the test specimen can be very large and some too large to be practical for both corrosion and fracture tests.

Having said that, for investigating the effect of corrosion on degradation of fracture toughness of structural steel, the primary purpose is (i) to experimentally observe how corrosion affects the fracture toughness of the steel but not actually to determine its accurate value of fracture toughness and (ii) to compare the corrosion effect on degradation of fracture toughness relative to its original value. It is, therefore, justifiable to select a smaller, practically manageable width for both corrosion and fracture toughness tests. This is because all test specimens should be under the same corrosion and fracture conditions, and hence, relative comparison of changes in fracture toughness over time with each other is valid. Besides, the accurate value of fracture toughness of structural steel can be determined with different methods as shown in, e.g., Gao et al (2020). A photo of fracture toughness set-up with the SENB specimen is shown in Figure 3.19 where the specimen is made according to ASTM E 1820 (2014). There is a procedure for testing and calculating the fracture toughness (K_{IC}) which is again provided in the relevant standards and beyond the scope of this book.

The results on fracture toughness reduction of specimens made of various mild steels immersed in a simulated soil solution (see Section 3.2.1) are shown in Figure 3.20,

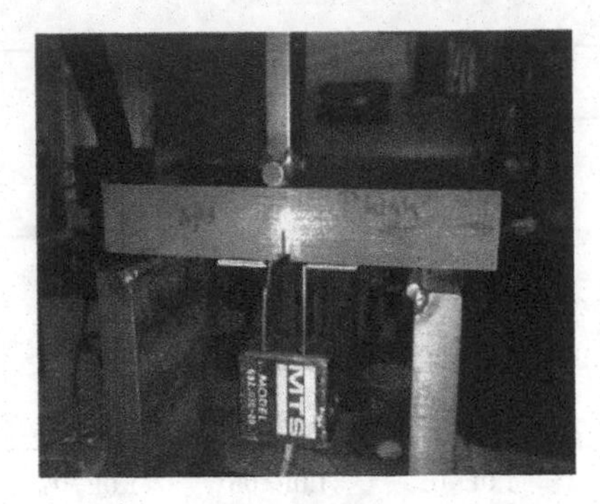

Figure 3.19 Test set-up for fracture toughness.

where each point represents an average of three testing results. The range of coefficients of variation of fracture toughness reduction at each point is from 0.12 to 0.17 over the test period. It can be seen from the figure that the fracture toughness decreases with time due to corrosion. This is true for different steels in different corrosive solutions. For example, in the simulated soil solution with pH = 5.5, a clear reduction of fracture toughness of Q235 steel can be seen from Figure 3.20a, which is 13.51% and 19.77% (from the original 189.47 MPam$^{0.5}$) after 180 and 270 days of immersion, respectively. Also, in the simulated soil solution with pH = 8.0, the reduction of fracture toughness of steel is 11.51% and 14.93% after 180 and 270 days of immersion, respectively. Higher acidity induced more corrosion and hence more reduction of fracture toughness. The results in Figure 3.20 provide the evidence that corrosion also affects the fracture toughness of the steel for the same reason as that explained for the tensile strength (see Section 3.5.2 for more details). In addition, the penetration of corrosion into the steel is not evenly distributed. In most cases, it is the locations that are damaged that incur the most corrosion, forming localised corrosion pits. This is true especially when there is a pre-crack where the corrosion is the most severe, leading to the extension of the crack of the specimen. As is known, the crack extension is one of the most important factors for the determination of fracture toughness (Li and Yang 2012).

Figure 3.20b shows the results of fracture toughness reduction of specimens made of G250 steel immersed in a different type of simulated soil solution (see Section 3.2.1), where each point represents an average of two close test results (Wang et al. 2019). These results indicate the same trend as for that of Q235 steel specimens. For example, in the simulated soil solution with pH = 2.5, a clear reduction of fracture toughness of steel can be seen from the figure which is 7.56% and 18.65% (from the original 170.45 MPam$^{0.5}$) after 180 and 365 days of immersion, respectively. This reduction is possibly due to the development of corrosion pits near the pre-crack tip (Rajani and Makar 2000), which facilitates crack extension, resulting in smaller values of fracture toughness. Similarly, the reduction of steel specimens in the simulated soil solution with pH = 5 is 2.25% and 12% after 180 and 365 days of immersion for the same reason as explained for that in the solution with pH = 2.5. A comparison of the results for fracture toughness of the same steel specimens in soil solutions with different pH values shows that a greater reduction is observed in specimens from soil solution with higher acidity over time than that in the solution with lower acidity. For example, after

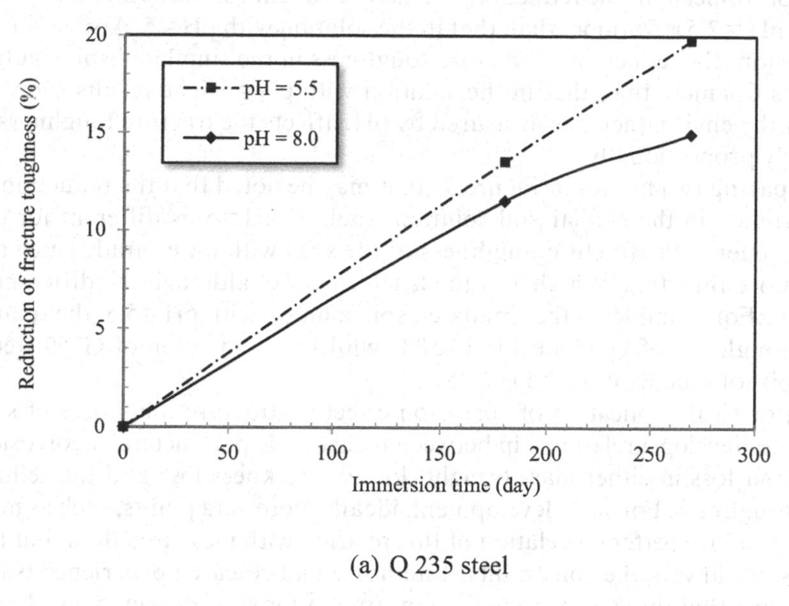

(a) Q 235 steel

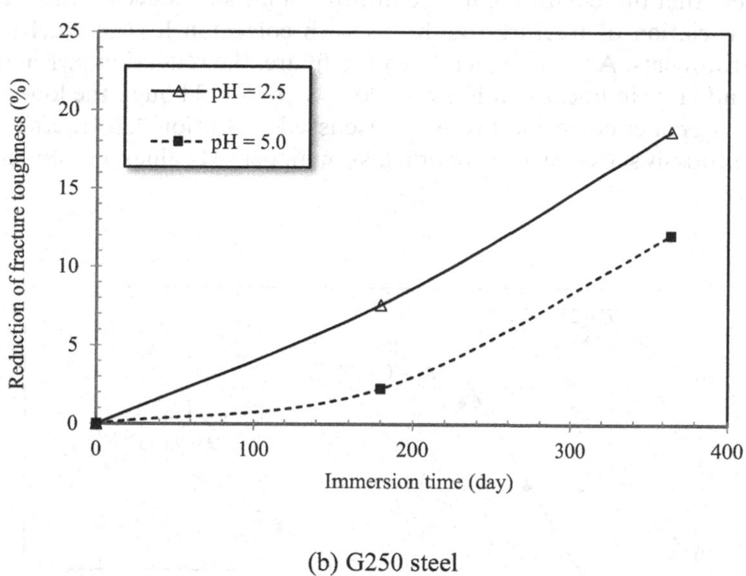

(b) G250 steel

Figure 3.20 Reduction of fracture toughness in simulated soil solutions for different steels: (a) Q235 and (b) G250.

180 days of immersion, the reduction of fracture toughness in the simulated soil solution with pH = 2.5 is 5% more than that in the solution with pH = 5. Also, after 365 days of immersion, the reduction of fracture toughness in the simulated soil solution with pH = 2.5 is 6% more than that in the solution with pH = 5. The results show that the acidity of the environment, as measured by pH, affects the fracture toughness of mild steel nearly proportionally.

Comparing two figures in Figure 3.20, it may be noted that the reduction of fracture toughness in the similar soil solution, such as pH = 5, is different for different steels. In general, the fracture toughness of the steel with lower grade (such as Q235) reduces more than that of higher grade (such as G250) although the difference is not significant. For example, in the simulated soil solution with pH = 5.5, the reduction of fracture toughness of Q234 steel is 13.51%, whilst the reduction of G250 steel in the similar soil solution with pH = 5 is 2.25%.

For practical application of corrosion effect on fracture toughness of steel, it is desirable to develop a relationship between measurable parameters of corrosion, such as corrosion loss in either mass (weight) loss or thickness loss and the reduction of fracture toughness. For such development, ideally more data points, such as more than four, can produce better correlation of this relation with measured data, but time and resources are always the constraints. Literature and research experience (such as Li 2001) suggest that three data points are minimum for such development. Figure 3.21 shows the variation of fracture toughness with corrosion loss of steel under three tested environments. As can be seen from the figure, the reduction of fracture toughness is by and large in linear relation with corrosion loss. Though the lower pH values contribute to greater corrosion losses as discussed in Section 2.2, fracture toughness seems to be equally sensitive to corrosion loss in higher pH values, i.e., larger corrosion

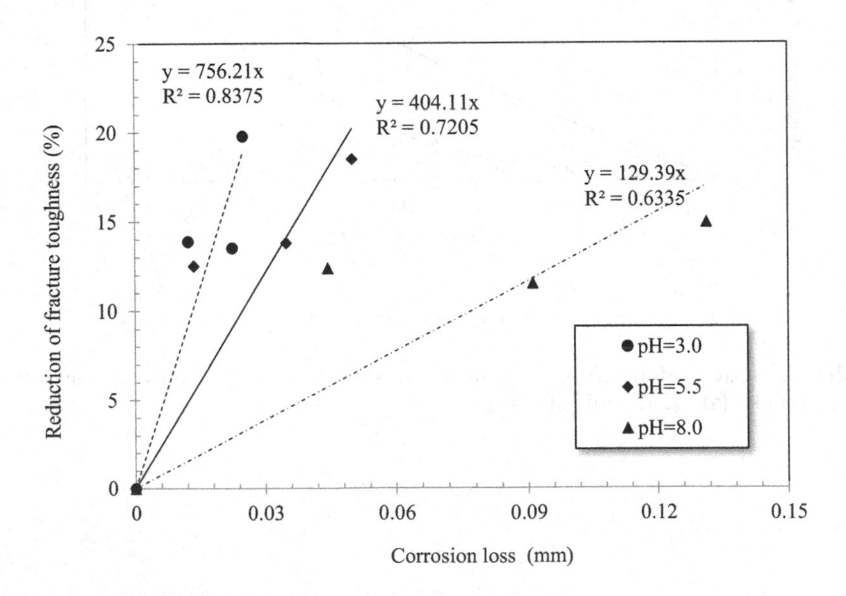

Figure 3.21 Reduction of fracture toughness with corrosion loss.

loss as shown in Figure 3.21. This can be because the acidity initiates more corrosion instantaneously, but reduction of mechanical property is a more accumulative effect of corrosion, such as corrosion loss. In practice, pH values of soil are rarely lower than 5. As such, results in Figure 3.21 for pH higher than 5 can be closer to reality and hence can be of more practical use.

It needs to be noted that the test results presented in the section are one step towards establishing understanding of and knowledge on corrosion effect on fracture toughness of steel. The significance of these results lies in its trend more qualitatively than the absolute value quantitatively. It is acknowledged that more tests are necessary to produce a larger pool of data for sensible quantitative analysis, based on which it is possible to develop theories and models for corrosion-induced degradation of fracture toughness of steel exposed in various aggressive environments.

3.4.3 Comparison of mechanical properties

It may be of interest to see how different mechanical properties deteriorate in the same corrosive environment. Such a comparison can be made from results of Sections 3.3 and 3.4. Table 3.8 presents the reduction of tensile strength and fracture toughness in the simulated soil solution shown in Table 3.1. It can be seen that the corrosion-induced degradation of fracture toughness is remarkably larger than that of tensile strength under the same conditions. This indicates that corrosion exerts a greater effect on fracture toughness than on tensile strength of the corroded steel. For example, after 90 days of immersion in the simulated soil solution, the reduction of fracture toughness is about three times larger than that of tensile strength. After 270 days of immersion in the soil solution, the reduction of fracture toughness is about 3–4 times larger than that of tensile strength. Also observed in the table is that the effect of acidity decreases after its pH is greater than 5. This effect is observed for both tensile strength and fracture toughness.

The reason for more reduction in fracture toughness can be that the pre-crack in the test specimens (as required for fracture toughness tests) promotes local corrosion, i.e., pitting corrosion at the crack tips further extends the existing pre-crack faster than otherwise no pitting. By definition, the extension of the crack marks the fracture toughness, which is actually reduced by the faster or earlier initiation of crack

Table 3.8 Comparisons of Reduction (in %) of Mechanical Properties of Steel

Exposure Period	*pH*	*Tensile Strength*	*Fracture Toughness*
90	3.0	3.43	9.29
	5.5	2.63	7.51
	8.0	1.39	6.35
180	3.0	4.66	14.22
	5.5	3.86	13.51
	8.0	3.31	11.51
270	3.0	4.77	19.77
	5.5	4.56	14.97
	8.0	4.43	14.93

extension. The results presented in this chapter are in consistence with those published in the literature. As an example, data from Garbatov et al. (2014) are taken for comparison, the results of which are shown in Figure 3.22. It can be seen that test results presented in the book are on a par with those published by Garbatov et al. (2014). Relatively, the results on reduction of tensile strength are closer than that of fracture toughness to those by Garbatov et al. (2014). The reasons could be that there are more uncertainties in determining the fracture toughness. Also, the time periods for two sets of data are quite different. Overall, the trend of two sets of data is the same or similar.

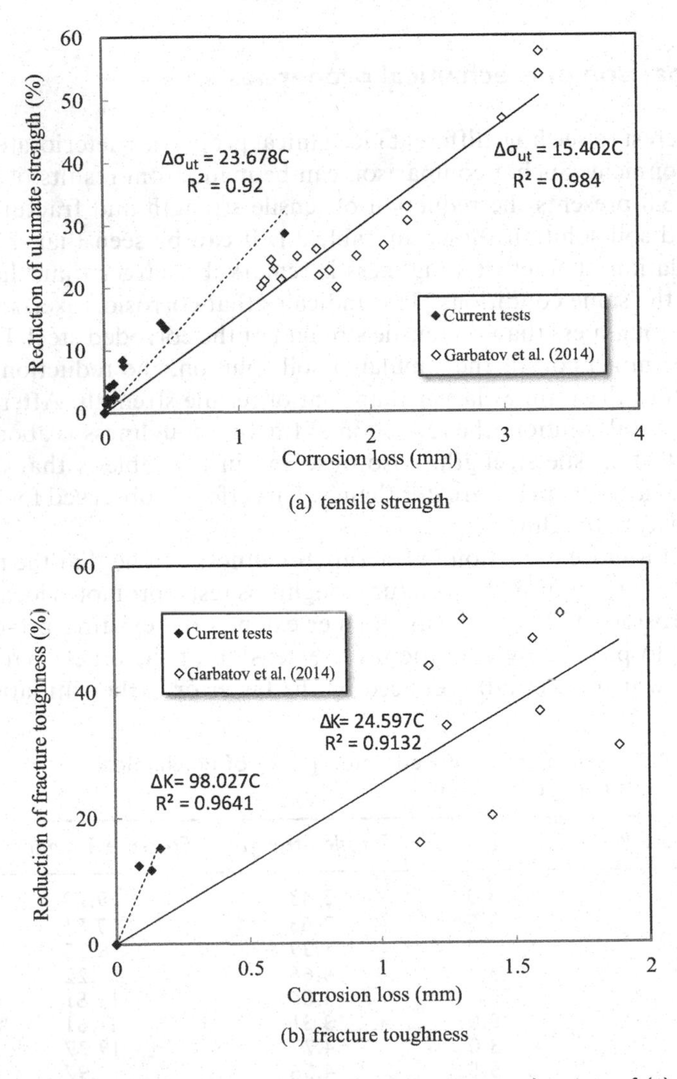

Figure 3.22 Comparison with data from the literature on reduction of (a) tensile strength and (b) fracture toughness.

As for corrosion effect on degradation of tensile and fatigue properties, it is possible to develop empirical formulae for reduction of fracture toughness. From Figures 3.21 and 3.22, the following empirical models can be developed from regression analysis:

$$\Delta K_{Ic} = 129.39C \text{ for corrosion in soil solution} \quad (3.16a)$$

$$\Delta K_{Ic} = 98.03C \text{ for corrosion in natural soil} \quad (3.16b)$$

where ΔK_{Ic} is relative reduction of fracture toughness in % (as defined in Equation (3.7)), and C is the thickness loss in mm. The coefficient of determination (R^2) of this regression is 0.91 and 0.96, respectively. It may be noted that the trend of reduction of fracture toughness in soil solution with pH = 8.0, which is almost natural soil environment, is very close to that in natural soil environment as can be seen with close coefficients of 129.39 and 98.03 in Equations (3.16).

3.5 MECHANISM FOR DEGRADATION

The mechanism of degradation of steel due to corrosion can be traced down to the atomic lattice of steel since the "golden rule" of material science is that the structure (atomic lattice) of a material determines the property of the material. A literature search can suggest that current state of the art in corrosion science and engineering has not developed a theory for degradation mechanism of steel at the atomic lattice level. From the perspective of engineering application, attention is more on changes of the microstructure of the steel that lead to the changes of its properties. This is the approach adopted in the section. It provides evidence from experimental observation and tests that the microstructure of steel has changed after corrosion. Changes of three important features of the microstructure of steel are presented in this section.

Firstly, the elemental composition of steel. Corrosion can consume the iron, reducing its content in steel during corrosion reaction. In the meantime, corrosion increases the oxygen content in steel due to the formation of ferrous oxides during the electrochemical reactions. In the meantime, iron content may be reduced during corrosion reaction. Also, salts in the environment, in particular, chloride ions, can enter the aqueous solution (electrolyte) and break the passive layer formed on the steel surface during corrosion which makes the steel vulnerable to pitting corrosion. In addition, hydrogen content can seriously affect the mechanical properties of steel. During the corrosion process, atomic hydrogen is released and accumulated at voids or defects within steel forming molecular hydrogen. The molecular hydrogen leads to inner pressure increase and microcrack initiations, which consequently degrades the mechanical properties of steel. The literature suggests that the elemental composition of steel affects its mechanical properties (Li 2018).

Secondly, changes of grain size. Corrosion can reduce the grain size primarily due to intergranular corrosion. The bonding force amongst grains can be weakened during intergranular corrosion, which degrades the mechanical properties of steel. Thirdly, changes in iron phases. As discussed in Chapter 2, iron contains two main phases in terms of its crystal structure, namely ferrite and pearlite. Ferrite is known as α-iron (α-Fe) which provides steel with strength and ductility. Pearlite is composed of ferrite (α-Fe) and cementite (Fe_3C) that makes steel brittle. Ferrite is corrosion-prone, and cementite is corrosion-resistant.

This section presents observations of changes in elemental composition, grain sizes and iron phases which are hypothesised as the mechanisms that lead to degradation of the mechanical properties of corroded steel as presented above. Again, this hypothesis is based on phenomenological observation rather than analytic derivation since the exact cause–effect relationship between the degradation and each factor or more factors remains a challenge to establish in particular the interactions amongst factors and resultant properties.

3.5.1 Changes in element composition

X-ray fluorescence (XRF) equipment, such as Bruker Axs S4 Pioneer XRF, can be used to determine the elemental compositions of steel. XRF is a technique that is commonly used for elemental analysis of substances with applications in the fields of science and engineering. XRF analysis is quick and does not require special preparation of the specimens. Samples from different exposure environments should be further cleaned with acetone and dried with air. They are then placed in the XRF equipment. The elemental compositions of the samples are measured and shown on the screen of the XRF equipment. Measuring locations are chosen as close to the steel/solution interface as possible since corrosion penetration depth is around 0.2 mm below the steel surface (Hu et al. 2011). Details on how to determine the elemental composition are beyond the scope of this book but can be referred to in Li et al. (2018a).

Selected results from the XRF tests for changes of iron, oxygen, chloride and manganese contents in steel specimens with exposure time are shown in Figure 3.23 for different exposure environments. As discussed in Chapter 2, these four elements are important to mechanical properties of the steel, and hence, their changes in content are hypothesised to affect the mechanical properties of the steel. There are also other elements that contribute to the mechanical properties of steel in one way or another. These cannot be covered one by one in the chapter due to the space limit of the book.

Figure 3.23 shows the changes of elements in HCl solution for various pH values. In Figure 3.23a, iron (Fe) decreases from 93.01% to 76.74% in HCl solution with pH = 5, 74.78% in the solution with pH = 2.5 and 58.60% in the solution with pH = 0 after 28 days of immersion. This reduction also explains the degradation of steel strength and ductility after corrosion. It is clear that the higher the acidity, the more the reduction. Simultaneously, oxygen (O) element increases from 5.92% to 20.8%, 22.78% and 38.27%, in the HCl solutions with pH = 5, 2.5 and 0, respectively. This is due to the formation of iron oxide (magnetite, lepidocrocite, etc.) during corrosion (Li et al. 2018a). Steel from the plant should contain very little chloride (Cl), but after corrosion, the chloride content increases to 0.24%, 0.39% and 1.36% in the HCl solutions with pH = 5, 2.5 and 0, respectively, after 28 days of immersion. The increase of chloride content is due to the penetration of chloride ions (Cl^-) into steel from immersion in the HCl solution and the formation of ferrous chloride ($FeCl_2$). Finally, Figure 3.23d shows that the manganese (Mn) content in the steel specimens decreases to 0.46%, 0.41 and 0.4% after 7, 14 and 28 days of immersion in the HCl solution with pH = 5. It decreases to 0.44%, 0.4% and 0.4% after 7, 14 and 28 days of immersion in the HCl solution with pH = 2.5,

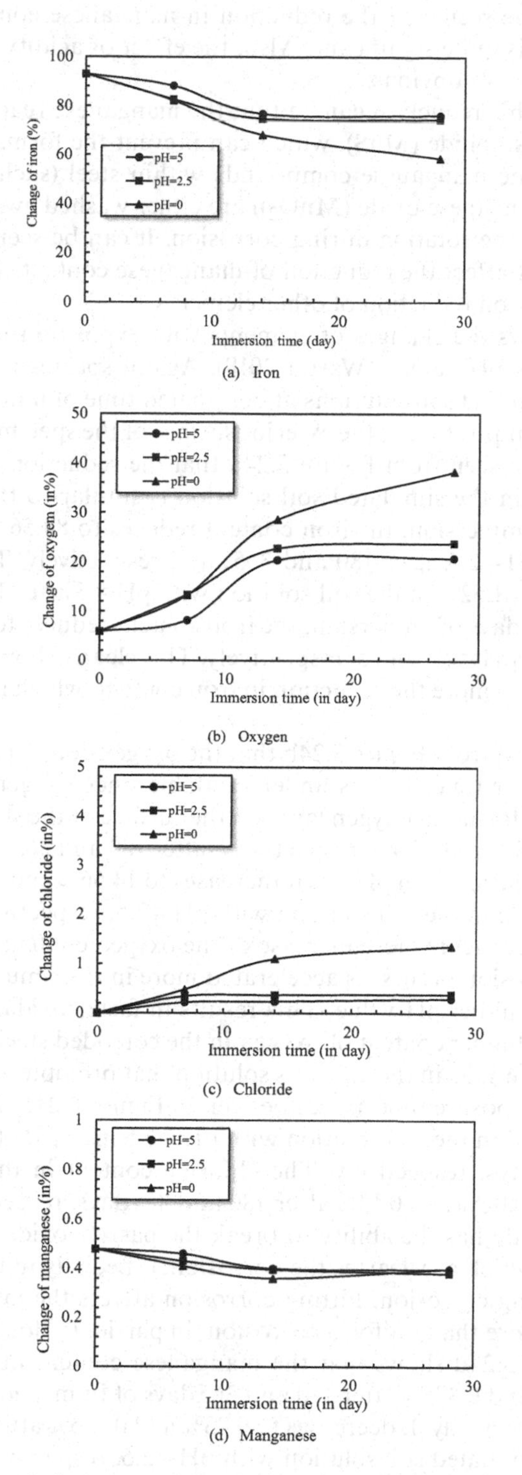

Figure 3.23 Reduction of element content with exposure time in HCl solutions.

respectively. It can be seen that the reduction in manganese content is in the range of 15%–20%, which is quite significant. Also, the effect of acidity on the reduction of manganese content is not obvious.

The reason for this reduction can be that the manganese reacts with sulphide (S) to form manganese sulphide (MnS), which can prevent the formation of ferrous sulphide (FeS). Also, the manganese compounds within steel (such as manganese sulphide (MnS) and manganese oxide (MnO_2)) are either washed away in the solution or reacted with HCl in the solution during corrosion. It can be seen that the acidity of the solution does not affect the reduction of manganese content significantly which is different to the effect on reduction of other elements.

Figure 3.24 shows the changes of elements with exposure time in simulated soil solutions for various pH values (Wasim 2018). Again, specimens are removed from their respective simulated soil solutions at designated time of immersion. After cleaning, the elemental composition of the exterior surface of the specimens was determined using XRF. It can be seen from Figure 3.24a that the reduction in iron (Fe) content with exposure time in the simulated soil solution is similar to that in HCl solution. Over the period of immersion, the iron content reduces to 84.56% and 66.56% in the soil solution with pH = 2.5 after 180 and 365 days, respectively. The iron content reduces to 91.19% and 81.52% in the soil solution with pH = 5 after 180 and 365 days, respectively. After 365 days of immersion, the iron content reduces to 66.56% and 81.52% in the solutions with pH = 2.5 and 5, respectively. This clearly shows that the higher the acidity (lower pH), the more the reduction in iron content, which is the same as in HCl solution.

It can be observed from Figure 3.24b that the oxygen content increases as the corrosion progresses over time. This is understandable since oxygen diffuses into steel, reacts with iron and forms an oxygen layer on the surface of the steel. The oxygen content increases to 5.58% and 15.4%, respectively, after 180 and 365 days of immersion in the simulated soil solution with pH = 5. It increases to 14.56% and 25.83% after 180 and 365 days of immersion in the soil solution with pH = 2.5, respectively. Again, it can be seen that high acidity incurs more increase of the oxygen content. The reason for this can be that the corrosion process is accelerated more in the simulated soil solution of higher acidity, i.e, smaller pH value. This results in faster oxidation reaction, which then accumulates a higher content of oxygen in the corroded steel.

Chloride ions are salts in the aqueous solution that promote corrosion. Its content increases with the exposure time as can be seen in Figure 3.24c. The chloride content in the steel immersed in the soil solution with pH = 2.5 increases to 0.257% and 0.37% after 180 and 365 days, respectively. The chloride content in the soil solution with pH = 5 increases to 0.06% and 0.13% after 180 and 365 days, respectively. As discussed in Chapter 2, chloride has the ability to break the passive oxide layer on the surface of steel specmens, which acceleartes the corrosion, especially in the form of localised corrosion, i.e., pitting corrosion. Pitting corrosion affects the mechanical properties of corrosion steel more than uniform corrosion, in particular for fracture toughness.

Finally, Figure 3.24d shows that the manganese content in the steel specimens decreases to 0.41% and 0.375% after 180 and 365 days of immersion in the soil solution with pH = 5 pH, respectively. It decreases to 0.43% and 0.375% after 180 and 365 days of immersion in the simulated soil solution with pH = 2.5, respectively. Again, the effect of acidity on reduction of manganese content is not obvious.

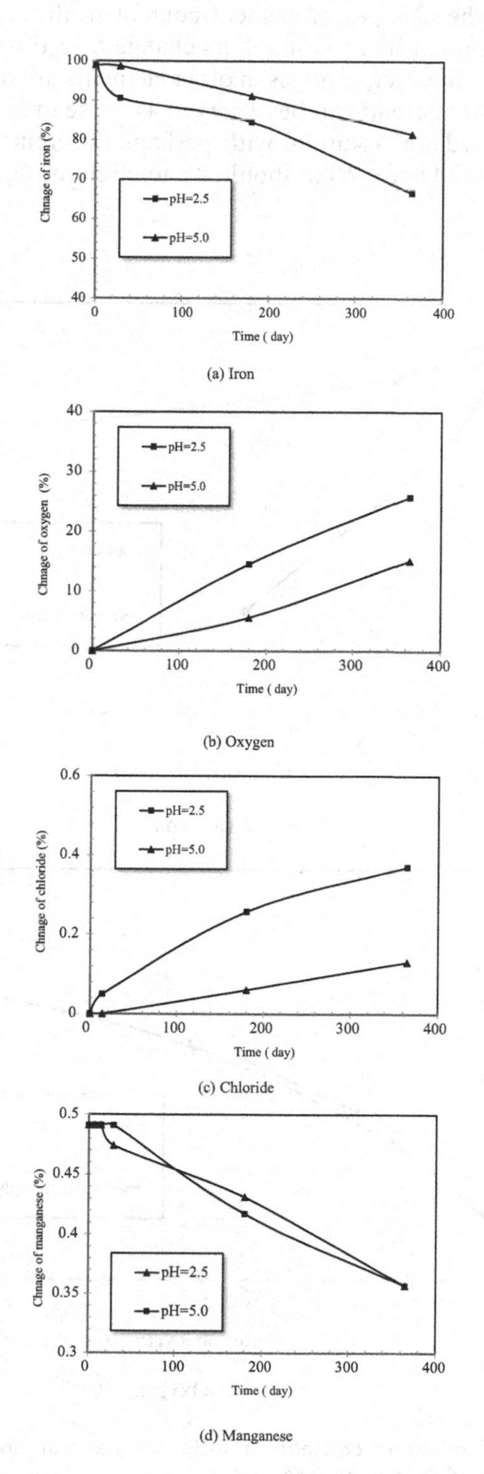

Figure 3.24 Changes of element content with exposure time in soil solutions.

Figure 3.25 shows the changes of element content in different natural environments. Since chloride content in air is small, its change over time is even smaller and hence can be negligible. However, changes in other elements are noticeable and shown in the figure. A very similar trend can be observed as those in acidic solutions, either HCl solution or simulated soil solution, with perhaps different percentage changes. Thus, it will not be repeated here. What should be noted in the figure is that the results

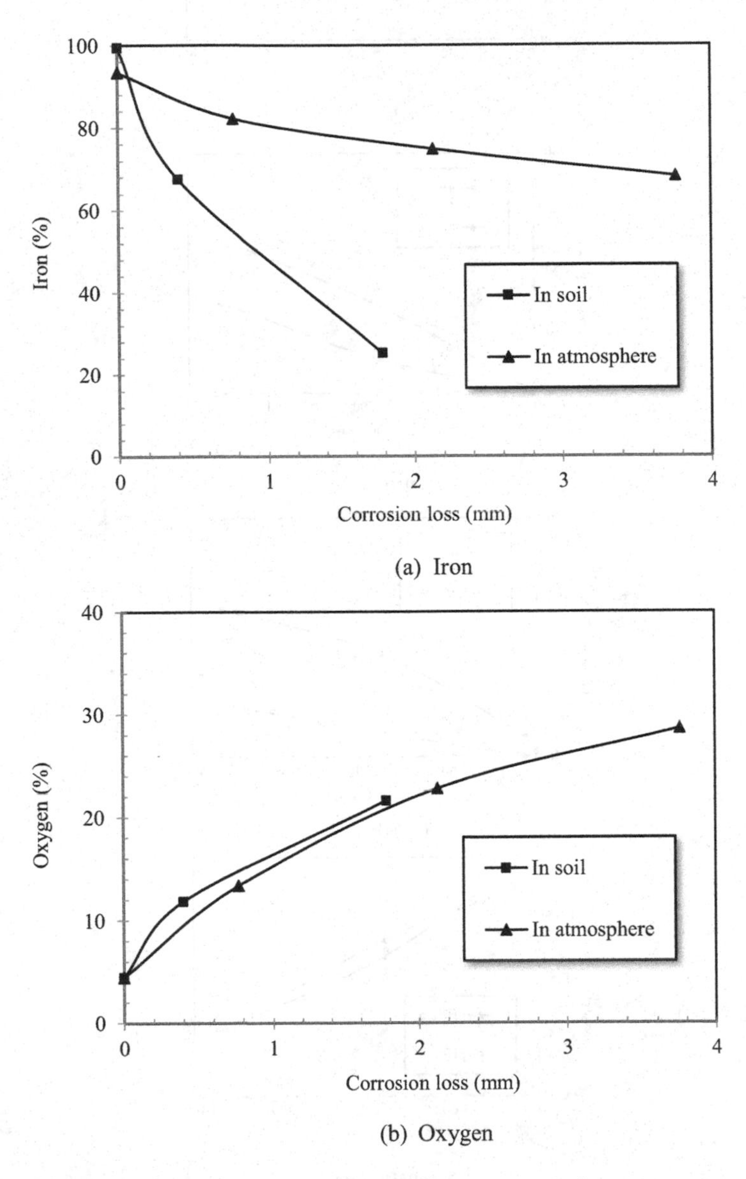

Figure 3.25 Changes of element content in different natural environments: (a) iron, (b) oxygen and (c) manganese.

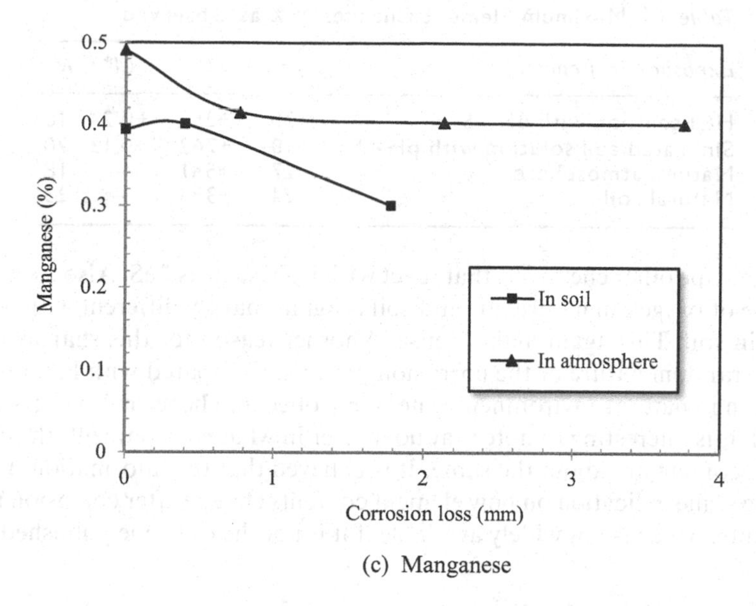

(c) Manganese

Figure 3.25 (Continued) Changes of element content in different natural environments: (a) iron, (b) oxygen and (c) manganese.

are from corrosion of steel in real service. This kind of information is rare as can be attested in the published literature. Also the significance of these results is the evidence that the chemical composition of steel has changed after corrosion. For example, after corrosion loss of 3.76mm, the iron content has changed from 93.36% to 68.14%, a reduction of 27.03% which is very significant and surely will change the properties of the corroded steel. Likewise, after corrosion loss of 3.76mm, the oxygen content has changed from 4.46% to 28.6%, an increase by six times. This is very significant, and intuitively, it will certainly degrade the mechanical properties of the corroded steel. Furthermore, Figure 3.25 shows that changes of elements in different environments are different. This indicates the randomness of the corrosion process and its effect on element changes.

What is of interest can be the comparison of element changes in different exposure environments. Since specimens are in different environments, such as solution and atmosphere, and also the changes are against different measurements, such as with exposure time and corrosion loss, a simplistic approach is taken to provide a qualitative indication of maximum changes of element contents. It is acknowledged that it is not ideal but does provide some indication and evidence that the content of different elements of steel does change with corrosion with different levels in different environments. The maximum changes of element contents in % as observed in different environments are summarised in Table 3.9, where "−" is decrease and "+" is increase. It can be seen from the table that in the acidic solutions, the changes of element contents are more or less the same, which are sharply different to those in natural environments. Even in the natural atmosphere and soil, the changes are sharply different. For example, the reduction of iron content is 27% in atmosphere and 74% in soil. This makes sense since in the

Table 3.9 Maximum Element Changes in % as Observed

Exposure Environment	*Fe*	*O*	*Cl**	*Mn*
HCl solution with pH = 5	−20	+251	+0.24	18
Simulated soil solution with pH = 5	−18	+262	+0.13	20
Natural atmosphere	−27	+541	–	18
Natural soil	−74	+384	–	23

soil, there may be other chemicals that react with iron, such as FeS. Also as can be seen, the increase of oxygen in natural air and soil is again sharply different with 541% in air and 382% in soil. This again makes sense. Another reason for this sharply difference may be the random nature of the corrosion process. Compared with laboratory tests, the corrosion in natural environment is not controlled and hence more random. On the other hand, it is interesting to note that no matter in what environment, the changes of manganese content are about the same. It is believed that the information in Table 3.9 can provide some indication on how element contents change after corrosion in various environments, which is not widely available if it is not the only one published.

3.5.2 Changes in grain size

Grain size can be quantified using an optical microscope (OM) at 100× magnification. Measurement can be conducted on samples of different exposure periods and environments. For grain size analysis, the cut samples need to be hot mounted with carbon and polished using silicon carbide grinding papers (180, 400, 600 and 1,200 grits), 3 μm diamond paste and 0.1 μm diamond paste. For grain size analysis, sample etching was required so prepared samples were etched with 2% Nital for 30 seconds (ASTM 2015a). Measuring locations are chosen as close to the steel/solution interface as possible since corrosion penetration depth is around 0.2 mm below the steel surface (Hu et al. 2011). A software ImageJ is needed to edit and analyse images (Li 2018). Usually, a linear intercept procedure is followed. The average grain size of steel used as specimens is 12.18 before corrosion. From Figure 3.26, it can be seen that after 28 days of immersion in HCl solutions, the grain size then reduces by 29.9%, 40.1% and 42.9% in the solutions with pH = 5, 2.5 and 0, respectively.

The OM images in Figure 3.27 confirm that grain size next to the steel/solution interface reduces as immersion time increases. The main mechanism for grain size reduction is intergranular corrosion (Sinyavskij et al. 2004). Intergranular corrosion occurs due to the difference in elemental composition between grain and grain boundaries. Subsequently, the reduction of grain size leads to degradation of mechanical properties by weakening the bonding force between grains (Shimada et al. 2002).

Intergranular corrosion is one of the crucial mechanisms of the corrosion-induced degradation of mechanical properties Li et al (2019b) It is therefore important to estimate the level of intergranular corrosion during corrosion by monitoring the changes of grain size. This has not been paid sufficient attention in the past as shown in the published literature. Based on the results of the grain size analysis as shown in Figure 3.26, there is as much as 42.9% reduction of grain size after corrosion, which indicates that the specimens are subjected to serious intergranular corrosion. The test results suggest

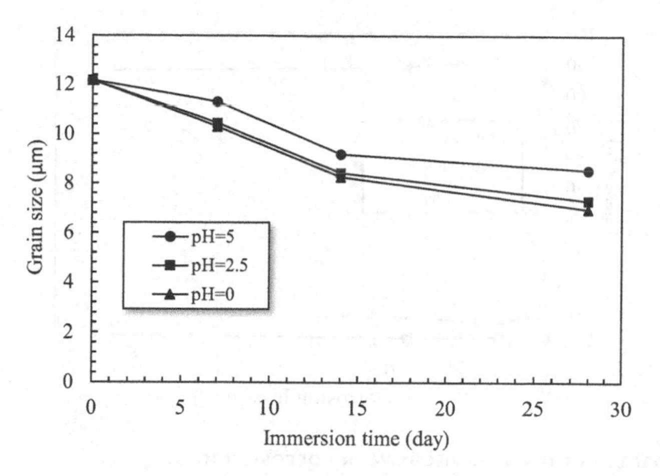

Figure 3.26 Reduction of grain size in various HCl solutions.

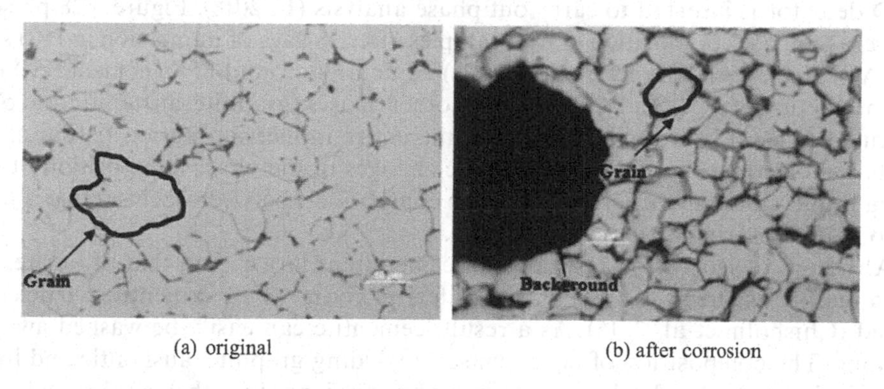

Figure 3.27 Reduction of grain size after immersion in HCl solution: (a) original and (b) after corrosion.

that protecting the steel from intergranular corrosion could be a way to prevent the degradation of mechanical properties. Also, in most of the literature, intergranular corrosion has been mainly focused on stainless steels, nickel base alloys, aluminum/magnesium alloys and mild steel in some particular solutions (such as Clark solution and $CQ_2 + NaNO_2$ solution) (Parkins 1994). Limited literature reports the concern of intergranular corrosion for mild steel immersed in acidic solutions, which is a common corrosive environment that the mild steel is exposed to. In this regard, test results in Figure 3.26 are very useful for both researchers and practitioners.

3.5.3 Changes in iron phase

Electron backscatter diffraction (EBSD) scanning, such as FEI Nova NanoSEM and Oxford Instruments Aztec software suite, can determine the average phase composition

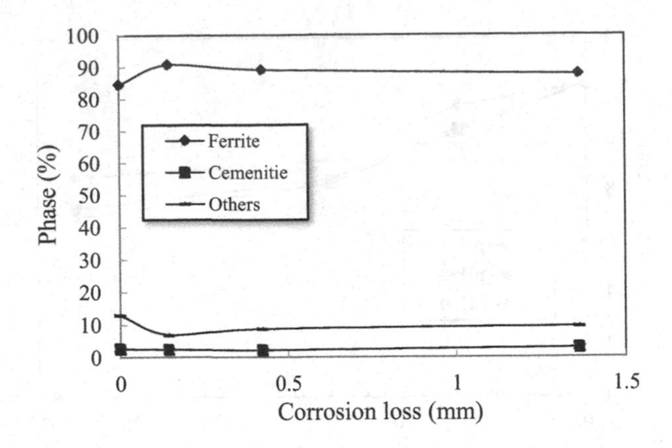

Figure 3.28 Changes of phase content with corrosion loss.

of iron. Samples taken out of the immersion need to be polished and loaded into FEI Nova NanoSEM to acquire its clear image at 1,000× magnification. Afterwards, the EBSD detector is inserted to carry out phase analysis (Li 2018). Figure 3.28 presents the average phase composition of iron samples after 28 days of immersion in HCl solution. As can be seen, the ferrite (iron bcc) content is around 85%, cementite (Fe_3C) content fluctuates at 2% and the content of other phases (graphite, austenitic and other undefined phases) is around 13% during the entire immersion period (28 days). The results indicate that there are no dramatic changes in the phase composition of iron during immersion. The difference in phase composition between each sample is likely due to manufacturing instead of corrosion.

Although the corrosion resistance of cementite is larger than that of ferrite, corrosion mainly occurs at the boundaries of ferrite grains where cementite particles are located (Chisholm et al. 2016). As a result, cementite can easily be washed away by solutions. The composition of other phases, including graphite, austenitic and impurities within steel, can also be washed away by solutions since they are located at the boundaries of ferrite grains (Chisholm et al. 2016). Consequently, the level of reduction of ferrite, cementite and others is similar, and there are no significant changes in their proportion. With the presence of stress, cementite is more likely to be fractured than ferrite (Umemoto et al. 2003). However, the intergranular stress corrosion cracking (IGSCC) also contributes to the corrosion of ferrite grains and other phases being washed away by solutions earlier (Arioka et al. 2006). As a result, the level of reduction of ferrite, cementite and others is still similar.

As a further discussion, it may be noted that depending on the depth of corrosion, i.e., the corrosion loss, the changes in microstructural features, i.e., element content and grain size, may only take place on the surface of steel and in particular not affect the bulk steel. The real question is how deep the corrosion penetration is needed to affect the property of bulk steel. This is a question without answer at the moment. However, as will be presented in Chapter 6, an idea will be proposed to connect the mechanical properties at the microstructural level to those at the macrostructural level, i.e., property of bulk steel.

3.6 SUMMARY

Observations of steel corrosion in three different environments, namely acidic solution, simulated soil solution and natural atmosphere are presented in chapter. Corrosion progress is expressed in time, in mass loss and in dimension loss. After corrosion, the test results of corrosion effects on tensile properties of steel are provided and discussed first, including degradation of yield strength, ultimate strength and failure strain of corroded steel in different environments. It is clear that all tensile properties degrade over time during corrosion, with ultimate strength degrading more than yield strength. It follows by the test results on degradation of fatigue strength and fracture toughness of corroded steel in different corrosive environments. It is clear that both degrade over time during corrosion. A comparison of corrosion effects on different mechanical properties in the same corrosive environments shows that fracture toughness degrades most in corrosion amongst all mechanical properties examined in the chapter, namely, yield strength, ultimate strength, failure strain, fatigue strength and fracture toughness. Then, the mechanisms of degradation of mechanical properties of steel due to corrosion are discussed based on the observations of microstructural changes in elemental composition, grain size and iron phase. It is clear that corrosion changes the microstructure of steel and subsequently the mechanical properties of steel. This information can be of great interest to practitioners and researchers alike.

Chapter 4

Corrosion impact on mechanical properties of cast iron and ductile iron

4.1 INTRODUCTION

As presented in Chapter 2, cast iron is a type of steel with more carbon content, usually larger than 2%. Cast iron also includes ductile iron, the difference from which is the shape of graphite, i.e., nodular or spheroidal graphite in ductile iron. Cast iron is perhaps the next mostly used metal to steel in construction and machinery although it is less used in above ground constructions, such as buildings and bridges, but more in underground constructions, in particular, pipelines. Some of old cast iron pipes are still in service although are being replaced once failed. Thus, the knowledge on corrosion of cast iron pipes could be very useful in terms of prolonging their service life and achieving sustainability. In general, the current knowledge of corrosion is more on steel than on cast iron. For the corrosion of cast iron, current literature focusses more on corrosion progress than its effect on the degradation of mechanical properties of cast iron, in particular, fracture toughness. It is the latter that determines the service life of corroded cast iron and cast iron structures, such as pipes. In addition, current tests on corrosion of cast iron are more in solution than soil and more material specimen (such as coupon) than structural specimen (such as pipe section). This chapter presents the effect of corrosion on the degradation of mechanical properties of cast iron primarily in soil environment with both coupon and section specimens.

There are two main modes of failures for cast iron (cast iron pipes): by rupture due to the reduction of dimension and tensile strength of the cast iron and by fracture due to the stress concentration at the tips of cracks, such as corrosion pits, or, in general, defects in the cast iron (Li and Yang 2012). The mechanical properties corresponding to these two failure modes are tensile strength, modulus of rupture and fracture toughness. Thus, this chapter focuses on the corrosion effect on the degradation of these three important mechanical properties of cast iron and ductile iron. The approach adopted in this chapter is again a mechanistic one based on phenomenological observations from laboratory experiments and field inspections. It is believed that this is a realistic and effective approach in dealing with corrosion effect on mechanical properties of corroded cast iron without losing the basic principles of the corrosion process, i.e., the electrochemical reactions. This approach can produce results that are directly related to cast iron and cast iron structures, such as pipes, to be designed and assessed.

There are many unique features of information and/or test data presented in this chapter compared with those of the published literature. The first one is the exposure environment. Three kinds of environments are adopted in the tests to produce the results on corrosion-induced degradation of mechanical properties. These are the commonly used acidic solution for immersion tests, the real soil to simulate the real burial tests and the natural soil in which cast iron is buried as the common practice for underground pipes. The second feature is the specimens used in the tests for both corrosion and mechanical tests. One is coupon, i.e., a piece of cast iron (material), as a specimen for both immersion tests and soil burial tests on corrosion. The other is a section of pipe (structure) as a specimen for soil burial tests on corrosion. Then a coupon, or a piece of cast iron, such as a plate of cast iron, is cut from the section to perform the same mechanical tests as performed on the coupon specimen. This provides the opportunity to compare corrosion of cast iron as a piece of material with that as part of a structure, such as a pipe. Such a comparison can provide useful information on testing methodology for corrosion of cast iron and indeed other metals where different environments are adopted. It can provide some kind of calibration of test results from different test specimens and environments. The third unique feature is the availability of exhumed cast iron pipes, which provides long-term natural corrosion of cast iron buried in soil with different conditions. It is believed that this information is very scarce, which makes this chapter and book unique.

It needs to be noted that although the chapter focuses on corrosion of cast iron and ductile iron in soil environments, the results can be applied to or at least referenced for corrosion in other similar environments, such as similar temperature, humidity and/or acidity. Also should be noted is that ductile iron is a type of cast iron. Thus, sometimes only cast iron is used, but it may mean or include ductile iron.

4.2 OBSERVATION OF CORROSION OF CAST IRON

The corrosion of cast iron is observed in simulated soil and in natural soil environments with test specimens as a coupon, such as a piece of cast iron plate, and as structural section, such as a segment of cast iron pipe. In simulated corrosion tests, acceleration is usually necessary when the purpose of corrosion test is to determine its effect on the mechanical properties of cast iron since it takes longer time for corrosion to cause any degradation of mechanical properties. As stated in Chapter 3, the effect of corrosion on mechanical properties can be different under different accelerated corrosions. Some calibration of results from accelerated corrosion tests against those under natural corrosion is necessary for practical application. Also, a relative approach is adopted in examining the corrosion effect on mechanical properties for given conditions. As for steel corrosion, this approach does provide useful information on corrosion-induced degradation of mechanical properties of cast iron. Discussions on the applicability of results obtained to the design and assessment of cast iron structures, such as pipes, are provided in Chapter 6. Having said that, corrosion of cast iron in natural soil environments is also covered in this chapter although such cases or examples are not as many and as comprehensive as one would wish.

4.2.1 Simulated corrosion

Cast iron is mostly used for pipelines, which are buried in soil. Thus, to have practical relevance, the corrosion of cast iron is induced in the simulated soil solutions and real soil with controlled acidity and saturation, i.e., not natural soil.

4.2.1.1 Exposure environment

Corrosion of cast iron is an electrochemical reaction with the corrosive agents in the soil where it is buried. In order to represent this reaction in the laboratory, it is necessary to simulate the working environment of the buried cast iron. There are two methods to simulate the working environment; one is to bury the cast iron in a box of real soil, and the other is to immerse the cast iron in a solution that contains main chemical elements extracted from the real soil, known as simulated soil solution. Published literature suggests that most of current research employs simulated soil solutions for corrosion tests on cast iron in soil (Liu et al. 2009). Therefore, this section presents simulated soil solution first. One advantage of using soil solution is the ease to control the testing variables and also monitoring of corrosion behaviour.

The selected soil needs to represent the working condition of the cast iron. The chemical composition of the selected soil should be analysed in making the soil solution. The principle in simulating soil solution is that the key chemical elements of the soil sample and in soil solution are the same (Liu et al. 2009). An example of chemical composition in simulated soil solution against that in the real soil is presented in Table 4.1 (Hou et al 2016, Wasim 2018).

As corrosion of cast iron under natural soil conditions will take a long time to have any significant effect on its material properties, acceleration of corrosion is also necessary for cast iron and indeed for most corrosion tests (such as Li 2001, Mohebbi and Li 2011). Corrosion acceleration can be achieved by increasing the acidity of the soil solution, such as adding sulfuric acid (H_2SO_4). Usually, three degrees of acidity are selected to examine how the effect of corrosion on the mechanical properties of cast iron varies with acidity. Experience in corrosion tests and field inspection shows that the soil with $pH < 4$ is extremely corrosive for buried metals (Romanoff 1957, Petersen et al. 2013). This means that when the pH value of soil is 3.0, its induced corrosion can be significantly accelerated. On the other hand, the pH of natural soil is about 7.5–8.0. A middle value between them is 5.5. Thus, the simulated soil solutions can be made with pH values of 3.0, 5.5 and 8.0 to account for different pH values on corrosion effect on mechanical properties of cast iron. It is known that the added sulfuric acid may react with the chemicals in the simulated soil solution, but this reaction would happen in the same manner as with the soluble chemicals in natural soil (Yan et al. 2008, Liu et al. 2009). The point is that the pH of all solutions for immersion tests should be maintained the same and used as the measurement for the solution.

Another type of exposure environment for corrosion tests on cast iron is real soil. Since corrosion of cast iron in real natural soil can take a long time to have a significant effect on its mechanical properties, measures for acceleration are also necessary: one is to add hydrochloric acid (HCl) in soil, and the other is to control the moisture

Table 4.1 Example of Simulated Soil Solution

Soluble chemical content in real soil (wt %)	CaO 0.92	MgO 1.54	K_2O 2.17	Na_2O 0.60
Chemical content in simulated soil solution (wt %)	$CaCl_2 \cdot 2H_2O$ 0.036	$MgSO_4 \cdot 7H_2O$ 0.190	KCl 0.069	$NaHCO_3$ 0.540

content of the soil as these two factors are the most important variables that affect corrosion in soil as discussed in Chapter 2. In this exposure environment, soil is real but not natural since its acidity and saturation are changed to accelerate the corrosion. Thus, it is still called simulated corrosion.

Clay soil is one of the most corrosive soils where cast iron pipes are laid in practice. It is also widely available and as such can be ideal to be used for corrosion tests. The resistivity and pH of the soil should be measured before used (such as 23.46 Ωm and 8.17 for normal clay soil). In addition, both physical properties and chemical and mineral compositions need to be measured, an example of which is shown in Tables 4.2 and 4.3. Then, the soil is crushed to a uniform size to eliminate the effect of other factors on the variation in corrosion results, such as different aeration in soils with different particle sizes.

The soil used for the corrosion test of cast iron is prepared with three degrees of acidity, as measured by pH values and two levels of saturation as measured by moisture contents. One example of test soil is with pH = 2.5, 3.5 and 5 and with soil saturation of 40% and 80%, as measured by moisture contents of 10% and 20%. The pH of the soil should be maintained by adding HCl to keep the target pH at 2.5, 3.5 and 5. The saturation of soil also should be maintained by adding water to keep target saturation at 40% and 80%. Care needs to be taken to minimise the variation in aeration by keeping the soil void ratio constant (i.e., by controlling the uniform density of soil (1.6 g/cm^3)) for the entire depth of burial of the specimens. Since the temperature of the soil can affect the corrosion rate of cast iron as discussed in Chapter 2, the ambient temperature should be kept constant throughout the test. Humidity can also affect the saturation of soil and hence is better to be kept constant during the test. Key steps in soil preparation are outlined in Figure 4.1 (Wasim 2018).

4.2.1.2 Test specimens

There are many types of cast irons as discussed in Chapter 2, amongst which grey cast iron is perhaps the most widely used material for construction due at least partially to its low cost. Because of this, grey cast iron is often selected for corrosion research. Cast

Table 4.2 Physical Properties of Soil Sample

Property	*Liquid Limit*	*Plastic Limit*	*Plastic Index*	*Optimum Moisture Content*	*Target Dry Density*	*Specific Gravity*	*Resistivity*	*Original pH*
Value	29.1	20.4	8.71	14.62 (%)	1,600 kg/m^3	2.64	23.5 Ω m	8.17

Table 4.3 Chemical Composition of Soil Sample (mg/kg)

Chemical	SO_4^{2-}	NO_3^-	Cl^-	K^+	Na^+	Ca^{2+}	Mg^{2+}
Content	50.0	23.1	115.6	39.1	119.0	333.0	197.0

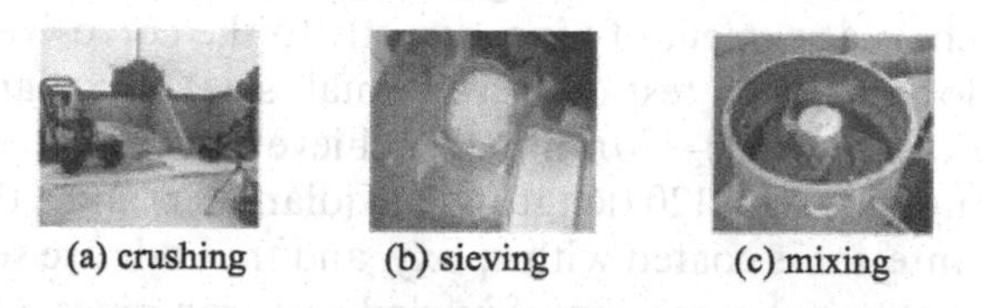

Figure 4.1 Preparation of test soil (a) crushing, (b) sieving and (c) mixing.

iron has quite different mechanical properties compared with those of steel due mainly to its brittleness. As it is well known, the mechanical properties of metal are affected by its chemical composition, morphology and microstructure, which vary significantly for different metals. In this section, two types of cast iron are presented. One is HT200 gray cast iron, which is widely used in the pipe industry in China (Hou 2016), and the other is T220 grey cast iron widely used in Australia (Wang 2018).

Two types of test specimens are prepared for corrosion tests and subsequent mechanical tests. One is coupon as material specimens and the other is a section of pipe segment as structural specimens for corrosion. For section specimen, a coupon is cut from the section for subsequent mechanical testing.

Specimens for corrosion tests are better made in the same shape and/or form as those for mechanical testing to avoid or minimise the damage during manufacturing of the specimens for different mechanical tests. This is particularly important for fracture toughness tests for which notches on SENB (single edge notched bending) specimens are difficult to make. Specimens for the corrosion effect on tensile strength presented in the section follow ASTM E8M13 Standard Test Methods for Tension Testing of Metallic Materials (ASTM 2013). Specimens for corrosion effect on fracture toughness test follow ASTM E1820-13 Standard Test Method for Measurement of Fracture Toughness (ASTM 2013). As presented in Chapter 3, in this standard, the key is to control the width of the specimen to achieve plane strain fracture. For immersion tests in simulated soil solution, specimens can be the same as those for steel corrosion tests for both tensile and fracture toughness tests.

For burial tests in soil, the dimensions and shapes of the specimens also follow the standard ASTM E1820 (2013). However, notch as required at the centre of the test specimen is made after corrosion tests or exposure. This is because when specimens are buried in the soil, corrosion may concentrate at the notches which will distort the real working condition to be simulated and subsequently affect the real mechanical properties of the otherwise less corroded specimens. This is quite different from immersion tests where corrosion is more or less uniform. Therefore, notches can be made on the corroded specimens after being removed from the soil environment. For practical use of the corrosion tests, the thickness is selected as that of real pipe wall which is 10–17 mm. With thickness of 10 mm, the dimensions of the specimens can be determined as 100 (length) × 20 (width) × 10 mm. Also, to simulate external corrosion of buried cast iron pipes, only one surface of the specimen is exposed to soil environment whilst other five surfaces or sides are coated with corrosion-resistant materials, such as epoxy, and wrapped with acid-resistant plastic to prevent the failure of the coating in acidic soil. Again, three replicate specimens are tested to allow for the variability of soil enjoinment and corrosion process.

Since cast iron is mostly used in underground pipes, there may be an advantage to expose the pipe, such as a segment of pipe, directly to the corrosive environment, i.e., buried in real soil for corrosion test due to its small sizes (the diameter of most cast iron pipes is in the range of 100–170 mm). To achieve this, a section or segment of a pipe is cut from a long pipe with 120 (length) × 120 (diameter) and a thickness of 11 mm. Each pipe section's interior is coated with epoxy, and the ends are sealed by polyethylene caps to simulate external corrosion of buried cast iron pipes. Once corrosion test finished, a plate is cut from the section wall to make a test specimen for mechanical testing following relevant standards.

4.2.1.3 Test procedure

Immersion corrosion test on cast iron is conducted according to a test standard, such *as ASTM G31-2012a Standard Guide for Laboratory Immersion Corrosion Testing of Metals (ASTM* 2012b). To examine the corrosion effect on the mechanical properties of cast iron over time, a minimum of three periods of exposure are selected in the tests, such as 90, 180 and 270 days. Before corrosion test, specimens need to be cleaned, such as using 50% acetone. After dry, they are placed in the containers of designated soil solutions with various pH values, such as pH = 3.0, 5.5 and 8.0. During the immersion tests, pH values of the solutions are checked constantly using a pH metre and adjusted by adding sulfuric acid to keep it constant during the duration of the immersion test.

To monitor the corrosion behaviour during the immersion tests, wires are welded to a control specimen in each container, and corrosion currents of the specimens are measured using an ampere metre. At the end of each designated time, such as the 90th, 180th and 270th day, specimens are removed from the solution for measurement of mass or weight loss and conducting mechanical testing. The detailed procedure of the immersion tests on cast iron can be found in Hou (2017).

In the burial soil tests, specimens are embedded in the prepared soil in an acid-resistant container where corrosion takes place and progresses. The soil is placed in the container layer by layer of 100 mm each to achieve a uniform density across the depth of the container. This is to eliminate variation in aeration in the pores of the soil. The burial depth of section specimens is 300 mm, based on field observations and reported literature (such as Goodman et al. 2013, Petersen and Melchers 2012). To eliminate the boundary influence, the distances between the specimen surface to both the walls and bottom of container are designed as 300 mm. To simulate the real underground environment, specimens are embedded on the bottom layer of soil of the container, serving as the bedding layer of the cast iron pipes. The test containers are partitioned with equal space to minimise disturbance to the soil and specimens whilst the specimens are removed out of the container for inspection, measurement and testing at the designated time.

During the period of corrosion test, the environmental conditions of the container are monitored by the moisture sensors, pH electrodes and thermocouples that are embedded in the soil. The water loss due to evaporation is measured by a high-capacity scale, and the water is replenished when necessary. Corrosion activities are monitored by the measurement of corrosion current, using linear polarisation resistance. The durations of burial soil tests are 210, 365 and 540 days. A schematic of the test set-ups

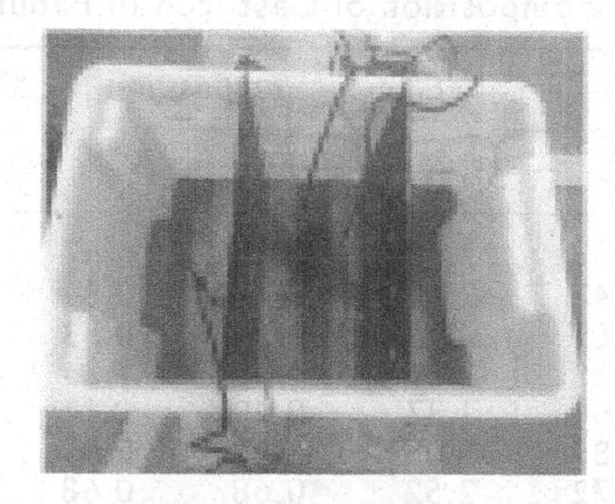

(a) Coupon embedded in soil

(b) Section embedded in soil

Figure 4.2 Set-up of burial soil test.

with all sensors are shown in Figure 4.2. More details of test set-ups and procedure for burial soil tests on cast iron can be found in Wasim (2018) and Wang et al. (2018).

4.2.2 Natural corrosion

Natural corrosion of cast iron and indeed for all metals may take a long time. As such, long-term corrosion tests on cast iron in natural soil are rare and only necessary for special circumstances. The only avenue to obtain corrosion information of cast iron in natural soil environment is from cast iron pipes exhumed from soil after decommissioning. It is very fortunate that such opportunity was seized from which exhumed pipes were brought to the laboratory for examination and measurement of corrosion and then mechanical testing.

Seven exhumed cast iron pipes are collected from three water utilities after they were exhumed from soil. The age of these pipes ranges from 37 to 79 years. The pipes are used for water distribution at around 50 m of water head. Five pipes have a diameter of 100 mm and two of 150 mm. The wall thickness of these pipes varies from 8.6 to 14 mm. The pipes are buried in soil with the burial depth varying from 0.8 to 1.5 m. To characterise the material types of the exhumed pipes, scanning electron microscopy (SEM) tests are first performed on the substrate of materials cut from the pipes. All specimens used in SEM and energy-dispersive X-ray spectroscopy (EDS) tests are prepared according to standards, such as ASTM E3-11 (2017). The morphologies of

Table 4.4 Chemical Composition of Cast Iron of Exhumed Pipes

Pipe No.	*Age*	*Element composition (wt %)*					
		C	*Si*	*Mn*	*P*	*S*	*Al*
1	58	3.58	2.48	0.74	0.67	0.06	<0.01
2	52	3.47	2.29	0.41	0.84	0.08	<0.01
3	56	3.59	2.43	0.34	0.53	0.05	<0.01
4	60	3.25	2.31	1.30	1.23	0.07	<0.01
5	52	3.60	1.37	0.49	0.12	0.07	<0.01
6	37	3.89	2.05	0.71	0.65	0.07	<0.01
7	79	3.72	2.52	0.68	0.68	0.07	<0.01

pipe materials are also characterised in accordance with standards, such as ASTM A247 (2004). To further examine the difference in intrinsic materials of pipes, drillings from the undamaged areas (substrate) of pipes are analysed by a Varian 730-ES Optical Emission Spectrometer. The results of element compositions for the pipes are presented in Table 4.4. It can be seen from the table that cast iron has a carbon content of over 2% and silicon of 1%~3% which is the common range of cast iron. As expected, there is a distinct difference in element compositions amongst pipes due to the different manufacturers or casting techniques.

After material characterisation of these pipes, a series of cast iron samples are cut from the selected parts of the pipes for various tests to be presented below. For SEM and EDS tests, small samples, approximately 10 × 10 × pipe wall thickness, are cut from pipes and mounted in Bakelite (mould). The corrosion pit depths of pipes are directly determined by 3D scanning of the external surface of pipes after the corrosion products are removed. For the tests on mechanical properties of cast iron or pipe sections, both tensile tests and crush ring tests, samples are carefully selected as much representative as the corrosion condition of the pipe. Then specimens are made for the tests on tensile strength of cast iron and the modulus of rupture of the sections. For tensile tests, dog bone-shaped specimens are used in accordance with standard ASTM E8 (ASTM 2015). Three duplicate specimens are prepared from each pipe. The tensile test specimens have a length of 230 mm and a width of 20 mm in the centre (gauge).

For crush ring tests, each section specimen has a length of 100 mm, which complies with the requirement that the length of pipe must exceed 50% of the pipe diameter (Seica and Packer 2004). Three duplicates are prepared from each pipe. For fracture toughness, samples are carefully selected so as to be representative of the corroded cast iron. Then specimens are made for fracture tests in accordance with a national standard, such as ASTM E 1820 (2013d). Three specimens are prepared from each pipe.

4.2.3 Corrosion measurement

The measurement of corrosion includes (i) corrosion current; (ii) corrosion loss, either mass or thickness loss; and (iii) corrosion pit depth. Corrosion current has long been used as a major indicator for the corrosion behaviour of metals. Corrosion current should be measured every day in the first week of the test and then weekly until the

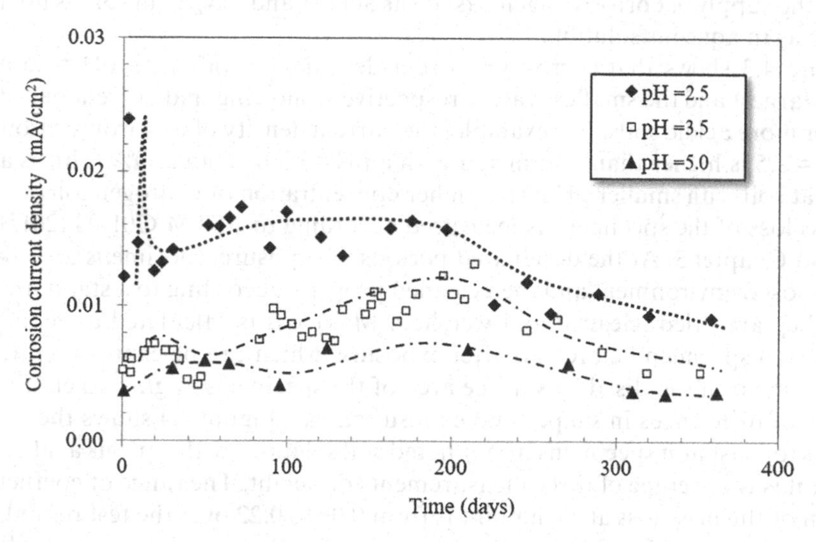

Figure 4.3 Corrosion current density of cast iron buried in various soils.

end of tests. An example of corrosion current of cast iron section in burial soil tests with various acidity values (pH) and 80% saturation is shown in Figure 4.3, where each point represents an average of three measurements with a standard deviation less than 0.003. It can be seen that, although each point of measured corrosion current is scattered, the general trend of corrosion currents is clear, which is decreasing with the exposure time. This means that corrosion rate is high at the beginning of the corrosion and decreases over time. As is well known, corrosion is an electrochemical process. The acidic environment can initiate the corrosion, but the progress of corrosion needs the supply of oxygen, which is not readily available to keep the high corrosion rate (Mohebbi and Li 2011). The figure also shows that corrosion currents are generally larger in a more acidic environment, i.e., smaller pH value, in particular at the beginning. Though corrosion currents for smaller pH are comparatively larger, the decreasing rates of corrosion currents (i.e., the slope of the curve) are irregular, exhibiting the randomness of corrosion behaviour.

It can be noted from Figure 4.3 that section specimens experience relatively high current densities at the beginning of the corrosion test (such as $i_{corr} > 0.02\,mA/cm^2$ for pH of 2.5) and the current densities gradually decrease with longer exposure time (such as $i_{corr} < 0.01\,mA/cm^2$ for pH = 3.5 and 5.0 after 250 days). This indicates that the corrosion rate is high at the initial exposure stage; however, it reduces and stabilises at a small value over time. This is not unexpected because in the early stage of corrosion, the section surface is completely exposed to a high concentration of hydrogen, iron and oxygen, which causes a high corrosion rate. In longer terms, an adherent layer of corrosion products forms a protective barrier against corrosion and consequently, the presence of a rust (oxide) layer slows or prevents the transportation of reactants (H^+ or O_2) to the iron substrate, resulting in a decreased corrosion rate (Hou et al. 2016, Mohebbi and Li 2011, Schwerdtfeger 1953). The decreased corrosion rate is also associated with the diffusion processes of corrosion reactants through the soil matrix because the supply of corrosive agents (such as acidity and oxygen in soil) is not readily available as in aqueous solution.

Figure 4.3 shows that corrosion current densities in soils with pH = 2.5 and 5.0 have the largest and the smallest values respectively, showing that current densities are larger for more acidic soils. For example, the current density of cast iron section in soil with pH = 2.5 is higher than that in soil with a pH = 3.5 by about 50%. This is a direct result that soil with smaller pH has a higher concentration of hydrogen ions.

Mass loss of the specimens is measured according to ASTM G01-03 (2017) as described in Chapter 3. At the designated periods of exposure, specimens are taken out of the exposed environment and then cleaned properly according to a standard. Afterwards, they are dried, cleaned and weighed. Mass loss is calculated as reduction in mass of each specimen before and after exposure, which can be expressed directly in mg with time or normalised by surface area of the specimens in g/m^2 to eliminate the influence of differences in shapes and exposure areas. Figure 4.4 shows the results of mass loss for cast iron specimens in simulated soil solutions with various acidity, where each point is the average of three measurements of weight. The range of coefficients of variation of the mass loss at each point is from 0.09 to 0.22 over the test period.

Figure 4.4 shows that the mass loss of cast iron increases with time almost linearly, which is different from the results of corrosion current (which is non-linear over time). The reason could be that the mass loss represents the cumulative effect of corrosion, which is more gradual, whilst the corrosion current represents the instantaneous rate of corrosion, which is more fluctuated. It can be seen that the mass loss is larger in a more acidic environment, i.e., the smaller pH value, such as pH = 3.0 in Figure 4.4. This is consistent with the results of corrosion current.

Corrosion rate is another commonly used measure for corrosion. It can be converted from mass according to ASTM G-01 (2014). It would be of interest to see how corrosion behaves in soil with various affecting factors. As described in Chapter 2, of

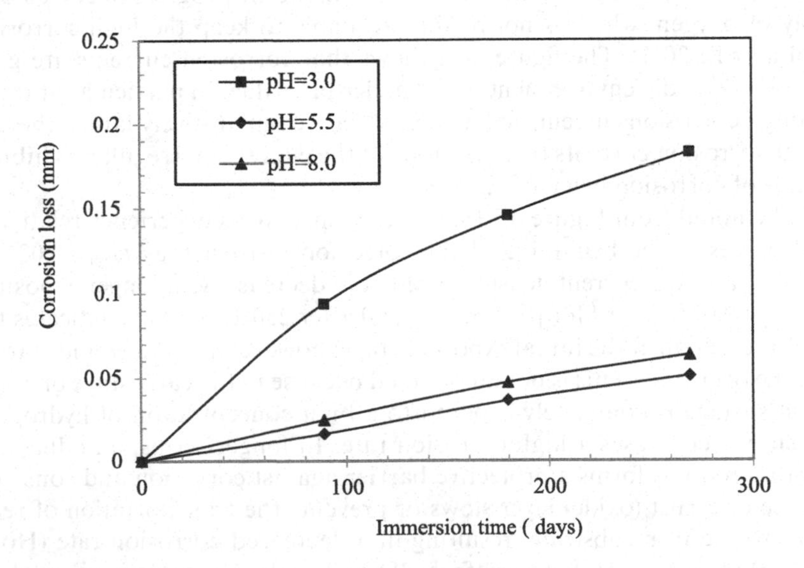

Figure 4.4 Mass loss of cast iron in simulated soil solution.

many such factors, acidity and saturation are the most significant factors that affect corrosion in soil environment. Corrosion rates of cast iron coupons buried in real soil (see Figure 4.2) for the following four typical combinations of soil environments: (i) high saturation with various acidity; (ii) low saturation with various acidity; (iii) high acidity with various saturation and (iv) low acidity with various saturation are shown in Figure 4.5 (Wasim et al. 2019b).

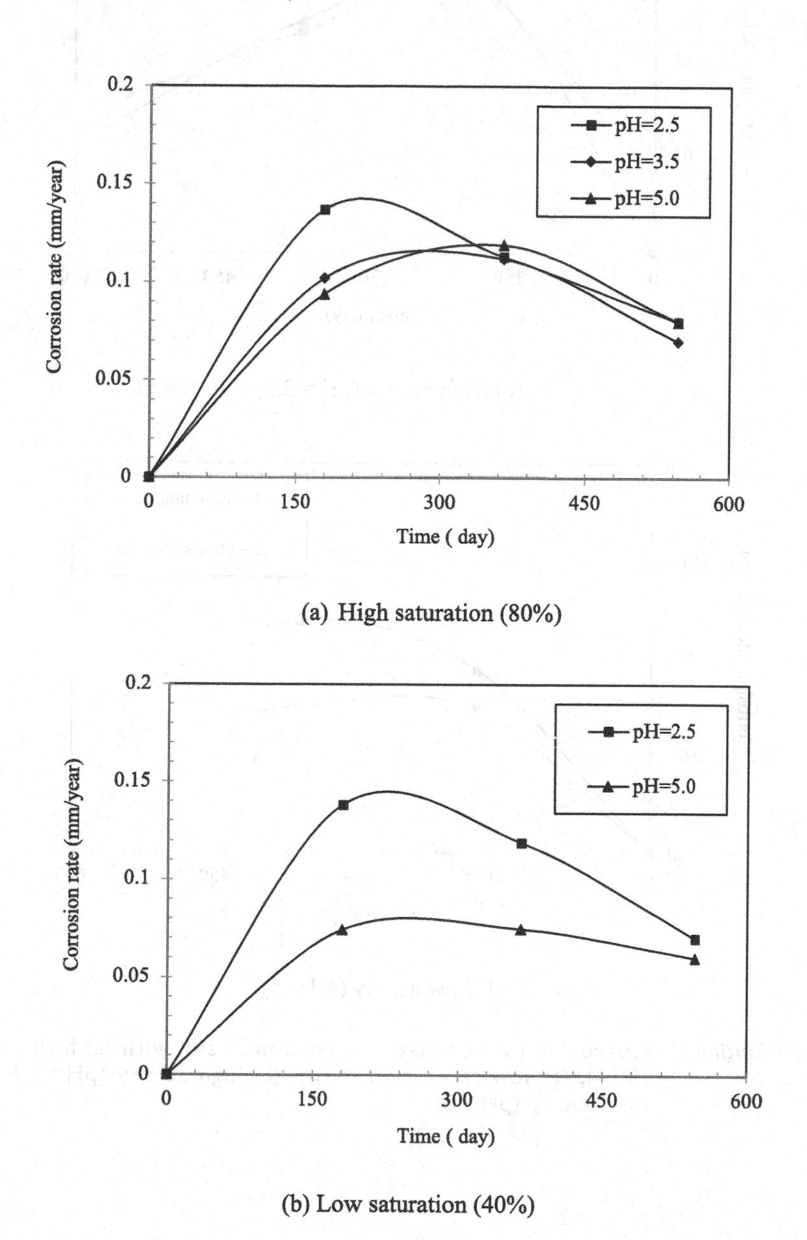

Figure 4.5 Corrosion rate of cast iron coupon in soil with (a) high saturation (80%), (b) low saturation (40%), (c) high acidity (pH = 2.5) and (d) low acidity (pH = 5).

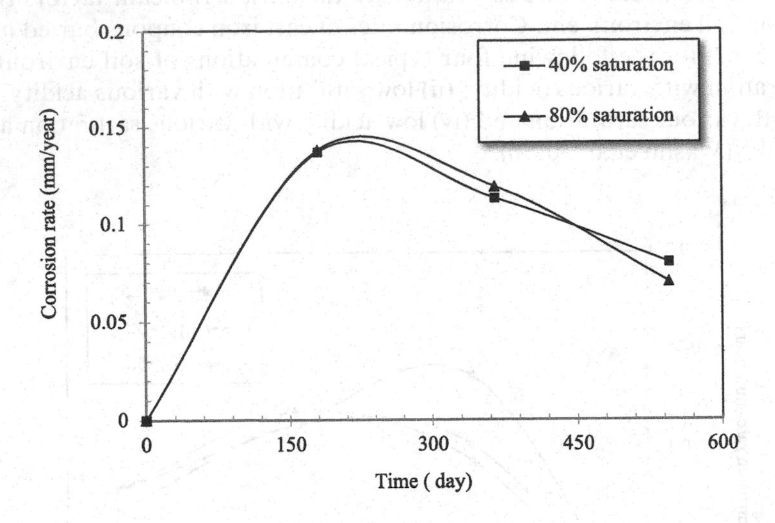

(c) High acidity (pH = 2.5)

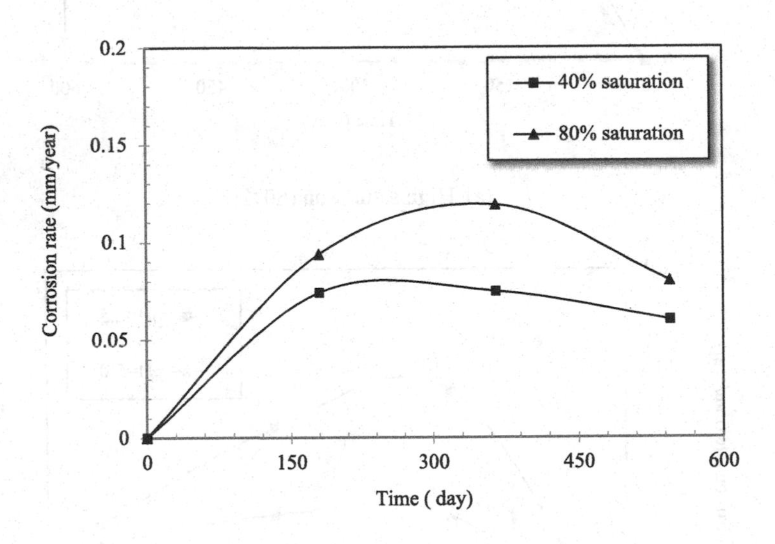

(d) Low acidity (pH = 5)

Figure 4.5 (Continued) Corrosion rate of cast iron coupon in soil with (a) high saturation (80%), (b) low saturation (40%), (c) high acidity (pH = 2.5) and (d) low acidity (pH = 5).

As can be seen, the highest corrosion rate (0.14 mm/year) occurs in soil with high acidity, i.e., pH = 2.5, and high saturation (80%) after 180 days. This result is expected since corrosion rate is usually high at the initial stage as conceptually depicted in Figure 2.3 (Section 2.2). As is known, a high saturation of 80% provides an easy path for dissolved oxygen through the pores of the soil to the specimens, leading to high corrosion rate. With the corrosion progress, rusts are developed on the surface of the specimens, which hinder further diffusion of oxygen and also access of moist to the surface of the specimens. This leads to the decrease of corrosion rate, such as 0.112 mm/year, at one year and smaller afterwards.

It can also be seen that corrosion rate is generally higher in soil with high acidity (smaller pH value) for given saturation as shown in Figure 4.5a and b. After one year, however, the trend is different. That is, in soil with high saturation, the corrosion rate becomes more or less the same or similar (0.119 mm/year) at one year, but in soil with low saturation, the corrosion of the specimens tends to be similar after about 1.5 years (540 days). It is clear from the figure that the soil with high saturation (80%) and high acidity (pH = 2.5) provides the worst corrosive environment for cast iron buried in it. On the other hand, the soil with low saturation (40%) and low acidity (pH = 5.0) provides the least corrosive environment for cast iron buried in it.

Overall, as the figure shows, the effect of acidity and saturation on corrosion rate is by and large the same (about 0.08 mm/year) in longer exposure time, such as after 1.5 years. This is understandable since with the corrosion products, i.e., rusts, accumulating on the surface, the supply of oxygen and water is more dependent on the diffusivity of the rust layer than the surrounding factors. The results presented in Figure 4.5 may have practical implication which is that in a longer term, corrosion progression depends more on the local environment, i.e., the electrolyte, and metal material itself, which produces the rusts that provide easy or difficult access of oxygen and water for corrosion to progress.

Figure 4.6 shows test results on corrosion rate for cast iron section specimens buried in soil with pH = 2.5, 3.5 and 5.0, and 80% saturation. As expected, although more

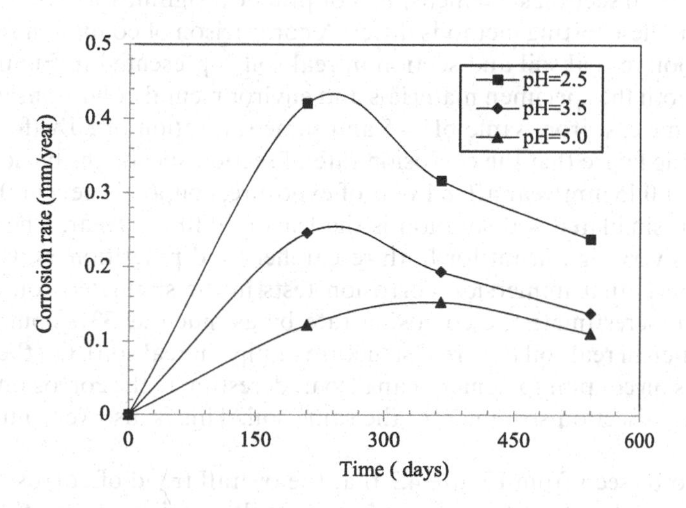

Figure 4.6 Corrosion rate of cast iron section in soil with various acidity and 80% saturation values.

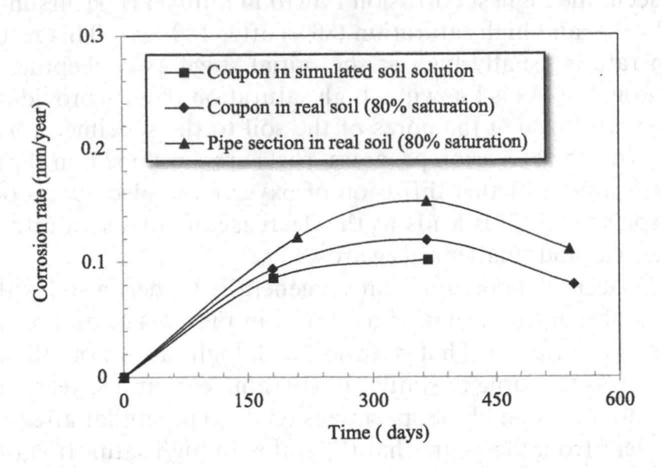

Figure 4.7 Comparison of corrosion rate produced with different methods: (a) coupon in simulated soil solution, (b) coupon in real soil and (c) section in real soil.

mass loss is caused with longer exposure times and lower pH levels, the corrosion rates generally decrease in all cases after a period of corrosion. For the example of the specimens buried in soil with pH = 2.5, the corrosion rate is 0.42 mm/year after 7 months of exposure and 0.32 mm/year at the end of 12 months. In comparison, the section specimens buried in soil with pH = 3.5 and 5.0 undergo less corrosion. The slopes of corrosion rates in both soils with pH = 2.5 and 3.5 indicate a reduction in corrosion rates over exposure time. The overall pattern for corrosion rate is in line with the conceptual model described in Figure 2.3 (Section 2.2).

At the moment, there are three different methods used for corrosion tests in terms of specimens and exposure environment. These include acidic solution, real soil, coupon specimens and section specimens. It is of practical significance to see how results produced from these testing methods differ. A comparison of coupon in simulated soil solution, coupon in real soil and section in real soil is presented in Figure 4.7. In the comparison, both the specimen materials and environmental conditions are the same, such as the same cast iron, same pH = 5 and same saturation of 80% (for soil). It can be seen from the figure that the corrosion rate of section specimens buried in real soil is the highest at 0.15 mm/year after 1 year of exposure, coupon is next at 0.12 mm/year and coupon in simulated soil solution is the least at 0.10 mm/year. The information in the figure is very significant for both researchers and practitioners. On one hand, it clearly suggests that immersion corrosion tests in the simulated soil solution can significantly underestimate the corrosion rate by as much as 33% compared with a section specimen in real soil (i.e., real structure or pipe in real soil). On the other hand, corrosion tests on coupon specimens can also underestimate the corrosion rate by 20% compared with a section specimen in the same soil. This is also very interesting and significant.

It can also be seen from Figure 4.7 that the overall trend of corrosion rate in all environments and by all testing methods are similar, which follows the conceptual model of Figure 2.3 (see Section 2.2). Since in most current corrosion tests on cast iron in soil, a coupon specimen, i.e., a piece of cast iron, and the simulated soil solution are used, the accuracy of the results produced from such tests needs very careful consideration twice. On a positive note, the results may provide some indication of some kind of correlation between different testing methods, using real structural (section) specimens in real soil as a benchmark. From this point of view, results can help both researchers and practitioners to understand the difference in corrosion rate by different testing methods and furthermore to correlate results from different methods.

One significant feature for corrosion of cast iron that distinguishes from other type of corrosion is pitting corrosion. This is because cast iron is mostly used for underground pipes where inhomogeneity of soil environment plus less uniformity of cast iron (compared with steel) creates local corrosion. Although the corrosion rates obtained from mass loss give an indication of the sectional loss of metals, localised corrosion in the form of pits in cast iron is a matter of primary concern, as it may result in fracture failure. Therefore, for practical application too, such as the pipeline industry, finding the maximum pit depth is imperative for the assessment of corrosion damage to structures (pipes) and for the estimation of their remaining service life. Pit depths of the test specimens can be measured using a 3-D profilometer, such as ContourGT-Ksystems 3-D profilometer, at the end corrosion exposure.

An examination of all tested specimens found that specimens buried in the soil with low acidity (pH = 5.0) and high saturation (80%) exhibit a high density of pits with varying sizes. On the other hand, the specimens buried in the soil with high acidity (pH = 2.5) and the same saturation display smaller, evenly distributed pits in their exposed surface. One example of 3D surface topography and pit depth of specimens in soil with various acidity values and high saturation (80%) is shown in Figure 4.8 (Wasim et al. 2020a).

It can be seen that the largest pit depth (0.45 mm in Figure 4.8c) occurs in soil with low acidity (pH = 5.0) and high saturation (80%). The largest pit depth in the soil of high acidity (pH = 2.5) and the same saturation is 0.24 mm (Figure 4.8a). The results suggest that if the specimens continue to corrode at the same rate (0.45 mm in 1.5 years), there would be a pinhole in the specimen (with thickness of 10 mm) after 33 years $\left(=1.5\times\frac{10}{0.45}\right)$, which could be half the designed life of cast iron pipes (assumed 80–100 years). Hence, the coupled effect of low acidity (5 pH) and high saturation (80%) of clay soil can cause severe localised corrosion of cast iron pipes buried in it. To further

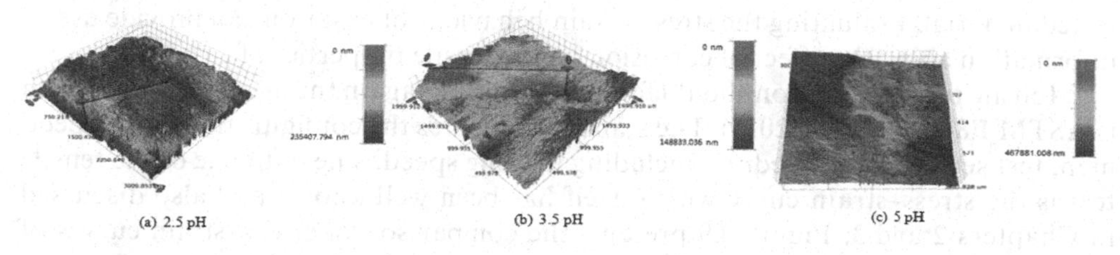

Figure 4.8 3D surface topography and pit depth of specimens in soil with varying acidity values and 80% saturation.

highlight the significance of the results of pitting corrosion, the deepest pit depth of 0.45 mm is five times larger than the corrosion rates (0.09 mm/year) of the specimens in the same soil environment, i.e., acidity with pH =5 and saturation of 80%. Clearly, corrosion rates which imply uniform corrosion are not the best indicator of risk of cast iron damage due to corrosion. The results presented in Figure 4.8 vindicate the need and significance to investigate the corrosion pit depth of cast iron buried in soil, perhaps more significantly in natural soil.

It may be noted that, as for almost all measurements in corrosion tests, variation of results is expected due to the stochastic nature of corrosion and measurement errors. The measurement errors from corrosion tests can occur during the process of, such as cleaning, weighing and making specimens. However, in theory, this kind of error should not happen. Overall, results in the figures presented in the chapter and the whole book are reasonably consistent and in reasonable agreement with those in published literature (such as Murray and Moran 1989, Romanoff 1957).

4.3 DEGRADATION OF MECHANICAL PROPERTIES OF CAST IRON

In principle, the corrosion process of cast iron is the same as that of steel, i.e., electrochemical reactions in the presence of oxygen and water. However, the corrosion effect on cast iron is quite different to that on steel, not just for material damage, such as more pitting, but also for degradation of mechanical properties, in particular the fracture toughness. In general, cast iron is a brittle material which means that its stress–strain curve does not exhibit clear yielding phase but ruptures suddenly under tensile stress. Therefore, there is no clear yield strength for cast iron. This is true in fact for most brittle materials. In addition, due to its brittle nature, cast iron is also not widely used in structures where cyclic loading is frequently encountered. As such, fatigue strength is not a trait of cast iron either. Although fracture toughness is also not the trait for cast iron, cast iron or cast-iron structure often fail by fracture. Thus, this chapter focuses on two practically important mechanical properties of cast iron: tensile strength and fracture toughness.

4.3.1 Reduction of tensile strength

Tensile strength is the very basic mechanical property of all metals, including cast iron. Tensile strength is tested on a material testing system, such as WAW-1000, and by qualified personnel, such as laboratory technicians, to ensure the quality of the test results. The direct outcome of a tensile test is the stress–strain relationship of the tested material. Evaluating the stress–strain behaviour of cast iron can provide useful information about the effect of corrosion on the tensile properties of cast iron.

Tensile tests of cast iron should follow a standard, again the mostly used of which is ASTM E8-15a (ASTM 2015). The standard specifies the configuration of the specimen, test set-up and procedure, including loading speed. The outcome of the tensile test is the stress–strain curve which itself has been well known and also discussed in Chapters 2 and 3. Figure 4.9 presents the comparison of stress–strain curves of uncorroded cast iron, the corroded cast iron with two levels of corrosion after buried in natural soils for 37 and 79 years respectively, i.e., from exhumed pipes. The

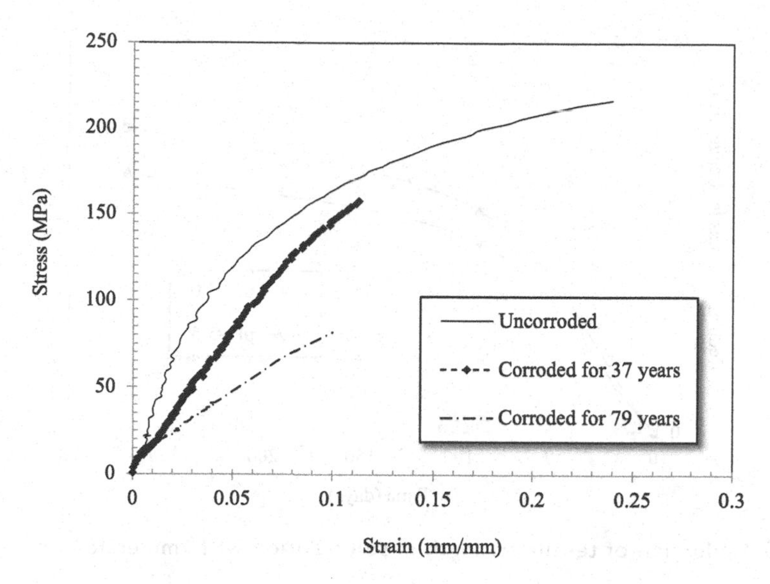

Figure 4.9 Stress–strain curves of cast iron with different levels of corrosion.

results in the figure show that the tensile strength of cast iron ranges from 140 to 200 MPa. Cast iron with low tensile strength could be the direct result that it contains relatively larger graphite flake size, such as larger than 0.2 mm in the edge area and 0.3 mm in the middle area of the corroded cast iron. As is known, larger graphite flakes in cast iron make it more brittle and easily break under tension. As expected, the specimens with surfaces free or less of any significant pits and graphitisation, such as uncorroded cast iron and corroded for 37 years, have higher tensile strengths than corroded specimens due to corrosion-induced material damage. Also, the degree of reduction of tensile strength depends on the extent of corrosion damage.

For the reduction of tensile strength, specimens are taken out of the exposure environments at designated times and then are cleaned according to a standard procedure as discussed before. After that, specimens are loaded to failure in tension on the testing machine. The results of tensile tests on specimens immersed in simulated soil solution for 90, 180 and 270 days are shown in Figure 4.10, where each point represents an average of three testing results. The range of coefficients of variation of tensile strength reduction at each point is from 0.11 to 0.19 over the test period.

From Figure 4.10, it can be seen that the tensile strength decreases with time due to corrosion. These results provide good evidence that corrosion does affect the mechanical property of cast iron as well. The main cause for this reduction may be that corrosion penetrates the surface of cast iron, destroying its compactness. Examinations of specimens taken out of the soil solutions reveal that surfaces of all corroded specimens are rougher and more porous than uncorroded, intact cast iron, which makes it easier for corrosive agents or other elements, such as O and Cl, to ingress into the metal. The ingress of corrosive agents and/or elements can alter the chemical composition of metal via chemical reactions of these agents and elements. It can also change the

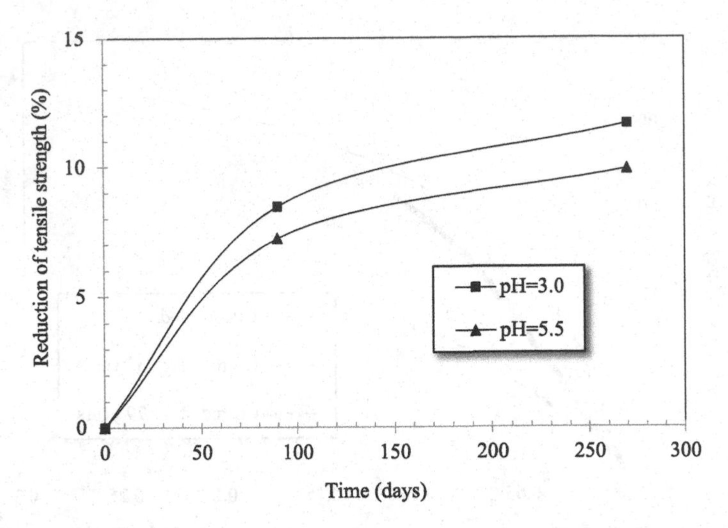

Figure 4.10 Reduction of tensile strength in soil solution with immersion time.

morphology or microstructure of cast iron. As is known, chemical composition and microstructure are the main factors that determine the mechanical property of metals. As a result, the mechanical property of cast iron changed.

It can also be seen from Figure 4.10 that the reduction of tensile strength of corroded cast iron increases with the exposure time. This is in line with that of corrosion loss (Figure 4.4), as discussed above. Compared with Figure 3.9 (Section 3.3), it can be seen that the reduction of tensile strength of cast iron is larger than that of steel. For example, the reduction of tensile strength of cast iron in soil solution with pH = 5.5 is 9.9% after 270 days of immersion, which is 6% for steel. This is again consistent with the results of both corrosion current and corrosion loss, indicating that high carbon content in metal may not only lead to more corrosion but also affect the tensile strength more. The figure shows that the reduction of tensile strength is larger for a more acidic environment (i.e., pH = 3.0).

The results of tensile tests on cast iron buried in natural soil, i.e., tests specimens cut from the exhumed cast iron pipes, are shown in Figure 4.11, where the corrosion loss is expressed relative to wall thickness as a general practice in the pipe industry. Also since pitting corrosion is more likely for cast iron, the reduction of fracture toughness is expressed in both uniform and pitting corrosion. It can be seen from the figure that no matter how corrosion is expressed, the tensile strength of cast iron reduces with corrosion. It is of interest to note that uniform corrosion contributes more to reduction of tensile strength than pitting corrosion. The reason could be that tensile tests are under uniform stress, but pitting corrosion creates more damage in non-uniform localised stress.

For practical application of test data on corrosion-induced reduction of tensile strength of cast iron buried in natural soil, it is desirable to develop an empirical relationship between measurable parameters of corrosion, such as corrosion loss, and the

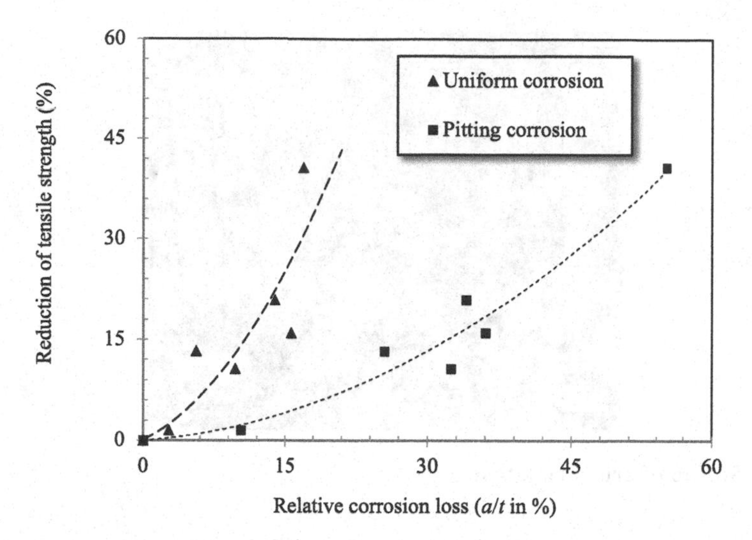

Figure 4.11 Reduction of tensile strength of cast iron in natural soil.

reduction of tensile strength. From Figure 4.11, the empirical relations can be derived from regression analysis as follows:

$$\Delta\sigma_t = 0.0648\left(\frac{a}{t}\right)^2 + 0.7028\left(\frac{a}{t}\right) \text{ for uniform corrosion} \tag{4.1}$$

$$\Delta\sigma_t = 0.0114\left(\frac{a}{t}\right)^2 + 0.1036\left(\frac{a}{t}\right) \text{ for pitting corrosion} \tag{4.2}$$

where $\Delta\sigma_t$ is the relative reduction of tensile strength (as defined by Equation (3.7)), a is the thickness or pit depth and t is the wall thickness (of a pipe or section). The coefficient of determination of the regression (R^2) is 0.7487 and 0.949, respectively.

4.3.2 Reduction of modulus of rupture

Cast iron is mostly used for underground pipes, the strength of which is usually determined by crush ring test. This is practical because the diameter of the pipes is usually small such as less than 170mm. Thus, it is not difficult to carry out tests on a segment of pipe as a prototype test. On the other hand, crush ring test is closer to the real working condition of pipes buried in soil, such as the vertical earth load, traffic loads, etc. above the pipe (Seica and Packer 2004). In crush ring test, the stress in the cast iron section or ring due to external loads is usually not directly considered (Wang 2018). Instead, modulus of rupture is often used as a measure of load carrying capacity of the section. In crush ring test, the specimen, i.e., a section of pipe, is placed on the testing machine and crushed to failure. Loading continues until the specimen completely ruptures (i.e., failure). The maximum load at the rupture of the section is the capacity

Figure 4.12 Set-up of crush ring test.

or strength of the section or ring, known as the modulus of rupture. A photo of crush ring test is shown in Figure 4.12.

The modulus of rupture can be determined as follows (Seica and Packer 2004):

$$\sigma = 954\frac{P_{\max}\left(I+\bar{t}\right)}{L\cdot\bar{t}^{2}} \tag{4.3}$$

where σ is the modulus of rupture (in MPa), $P_{\max}$ is the load at fracture of cast iron ring or section (in N), L is the mean length of the section (in mm), I is the mean internal diameter of the section (in mm) and $\bar{t}$ is the mean wall thickness of the section measured at the fracture position (in mm).

As described in Section 4.2.1, the specimens for crush ring tests are a 120-mm-long segment of pipe like a ring. After buried in real soil for a designated time, they are taken out for crush ring test. To ensure good contact between loading plate and ring specimen, a resin cushion of approximately 1 mm thick is laid on the bed and bottom of the loading plates of the testing machine. The head of the machine is then pressed down at the rate of 0.3 mm/minute until failure occurs. At appearance of each crack, its size and location are carefully noted. This procedure continues until the ring specimen completely ruptures (i.e., failure). After the tests, the wall thickness of the section or ring at the fracture location is measured by a digital caliper to the nearest 0.01 mm.

During the test, a crack appears first either at the top or bottom of the ring section at a load close to the maximum capacity or strength of the ring. With the increase of load, this crack extends immediately through the wall of the ring section and breaks the ring, i.e., rupture, which occurs at the top or bottom for all ring specimens. The modulus of rupture is determined from Equation (4.3) with maximum load from the tests. Results on modulus of rupture from the crush ring tests for cast iron sections buried in soil with pH = 2.5 and 80% saturation and natural soil are presented in Figure 4.13. It can be seen that for sections or rings buried in the soil with pH = 2.5 and 80% saturation, the reduction of the modulus of rupture of the section ring is small compared with that in natural soil. This can be due mainly to

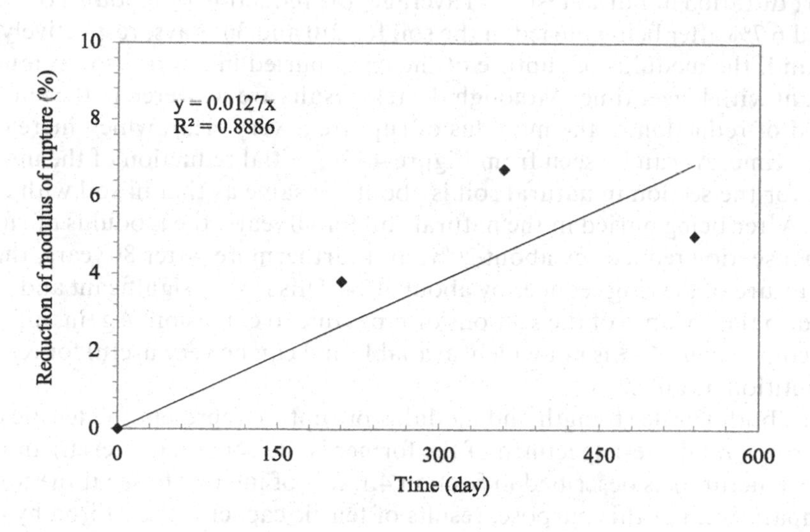

(a) Soil with pH = 2.5 and 80% saturation

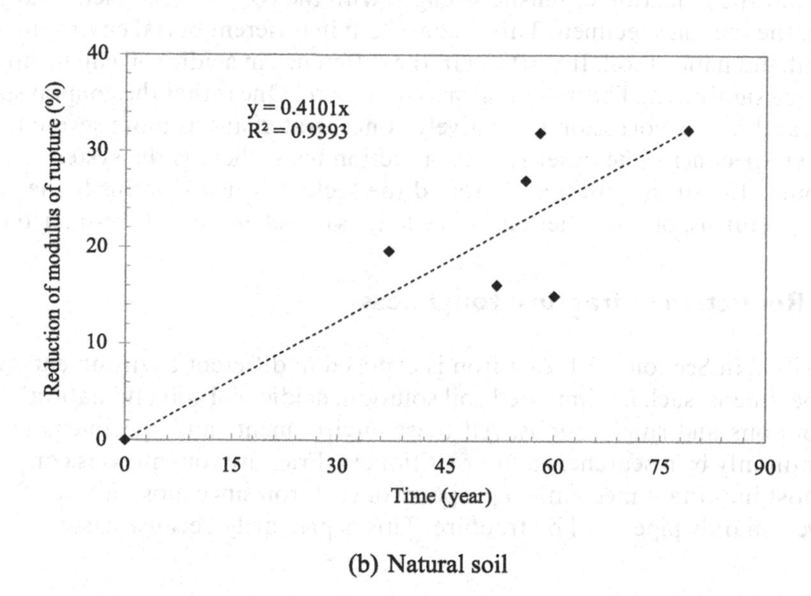

(b) Natural soil

Figure 4.13 Reduction of modulus of rupture in (a) soil with pH = 2.5 and 80% saturation and (b) natural soil.

the short duration of burial tests. On average, the reduction of modulus of rupture is 3.7% and 6.7% after being buried in the soil for 210 and 365 days, respectively. On the other hand, the modulus of rupture of the rings buried in natural soil is much more significant with longer time. Although the test results are scattered in the natural soil, the trend of reduction of the modulus of rupture is very clear, which increases with exposure time. As can be seen from Figure 4.13b, initial reduction of the modulus of rupture for the section in natural soil is about the same as that in soil with pH = 2.5, i.e., 6%. After being buried in the natural soil for 40 years, the modulus of rupture of the (pipe) section reduces by about 20%, and furthermore, after 80 years, the modulus of rupture of the ring reduces by about 30%. This is very significant and certainly will lead to the collapse of the sections or pipes due to corrosion. Again, information presented in Figure 4.13 is not widely available and can be very useful for researchers and practitioners alike.

Since both tensile strength and modulus of rupture represent the tensile capacity of cart iron, and the test specimen of the former is the coupon (material) and latter is section (structural) as described in Section 4.1, it is of interest to see if the test results are comparative. For this purpose, results of tensile capacity of cast iron by different specimens presented in Figures 4.10, 4.11 and 4.13 are summarised in Table 4.5. It can be seen that the reduction of tensile strength with the coupon specimen is larger than that with the section specimen. This is consistent in different burial environments, i.e., acidic soil and natural soil. In particular, the difference in acidic soil can be up to 42%. This is very significant. The main reasons can be two. One is that the coupon specimen is small, and hence, corrosion is relatively concentrated and is more severe than that of a larger specimen. The other is that in section tests, there is the system effect that redistributes the stress within and around the section which sustains higher load. In coupon specimens, on the other hand, the load is direct and hardly re-distributed.

4.3.3 Reduction of fracture toughness

As described in Section 4.2.1, cast iron is exposed in different environments with different specimens, such as simulated soil solution, acidic real soil and natural soil with small coupons and ring sections. All these environments and specimens have been used commonly by researchers and practitioners. Fracture toughness is considered to be the most important mechanical property of cast iron since most failure of cast iron structures, mainly pipes, fail by fracture. This is primarily because cast iron is brittle

Table 4.5 Difference in Tensile Capacity with Different Specimens

Environment	*Reduction in Tensile Capacity (in %)*		*Difference (in %)*
	Coupon	*Section*	
Acidic soil	10	7	42
Natural soil	40	32	25

and less homogeneous than steel for example. Any defects, such as corrosion pits, can lead to fracture due to stress concentration at the defects. For this reason, most of the literature on corrosion of cast iron is related to fracture. This section adds more information to the knowledge of reduction of fracture toughness of cast iron caused by corrosion in different environments and by different specimens.

One of the commonly used corrosion tests is immersion in simulated soil solution as presented in Section 4.2.1. After corrosion, the specimens are taken out of immersion and cleaned for fracture toughness tests, following a national testing standard, such as ASTM E 1820 (2013d). The results of fracture toughness reduction of cast iron in simulated soil solutions are shown in Figure 4.14, where again each point represents an average of three testing results. The range of coefficients of variation of fracture toughness reduction at each point is from 0.12 to 0.17 over the test period. It can be seen from the figure that the fracture toughness decreases with time due to corrosion. Initially, the effect of corrosion on reduction of fracture toughness is largely affected by the acidity of the soil solution in particular when the pH is smaller than 5.5, i.e., more acidic. In a longer term, however, the effect of acidity on the reduction of fracture toughness is less dependent on the acidity. The main reason for this phenomenon can be that high acidity induces more corrosion initially, and associated with it, there may be more corrosion pits for cast iron, which cause more reduction of fracture toughness. After a longer time, corrosion slows down as explained in Section 4.2.3, and corrosion progress is less dependent on the acidity. As such, its effect on reduction of fracture toughness is similar regardless of acidity. The results of Figure 4.14 again provide the evidence that corrosion does affect the mechanical property of cast iron for the same reason as explained for the tensile strength.

Table 4.6 shows a comparison of reduction in mechanical properties of cast iron in the same environments. It can be seen that initially, the effect of corrosion on tensile strength of cast iron is much smaller than that on fracture toughness, such as 7.22% reduction in tensile strength and 31.05% in fracture toughness, respectively, in the same soil solution with pH = 5.5. The main reason for this can be that corrosion pits extend the existing defects in the cast iron, which reduce the fracture toughness. In addition, examination of corroded cast iron suggests that the penetration of corrosion into the metal is not evenly distributed. In most cases, it is the locations that are damaged that incur the most corrosion, forming localised corrosion pits. In the longer term after 270 days of immersion, the reduction of tensile strength increases, whilst that for fracture toughness decreases, such as 9.89% and 32.79% reduction in tensile strength and fracture toughness, respectively. This may be because the initial cracks caused by corrosion pits do not continue to extend in the same proportion as initially. As it is known, the crack extension is the most affecting factor for the determination of fracture toughness (Li and Yang 2012).

One of the uniqueness of the test data presented in the chapter is that cast iron specimens are buried in real soil with various acidity (measured by pH) and saturation (measured by moisture content) values. At designated times of burial, specimens are taken out of the soil and cleaned. After corrosion measurement (see Section 4.2.3), tests on fracture toughness of corroded cast iron are conducted, following a national testing standard, such as ASTM E 1820 (2013d). Results on the fracture toughness of coupon in soil with various pH values and saturation are presented in Figure 4.15 as a function of exposure time, where each point represents the average of two test results

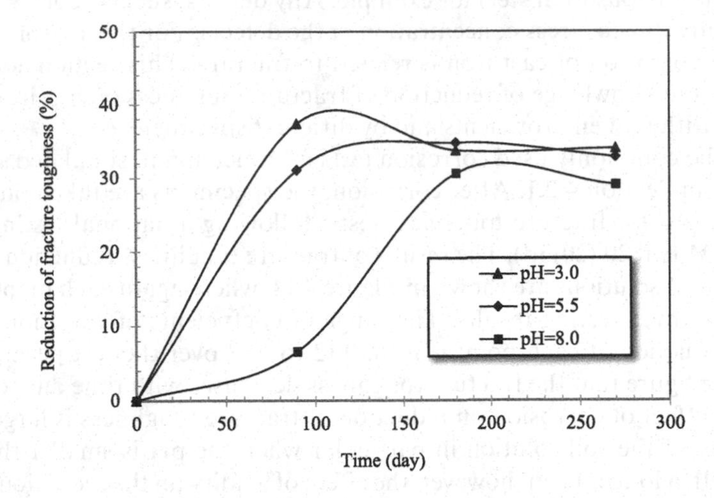

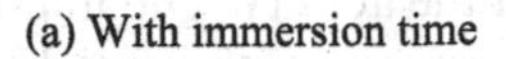
(a) With immersion time

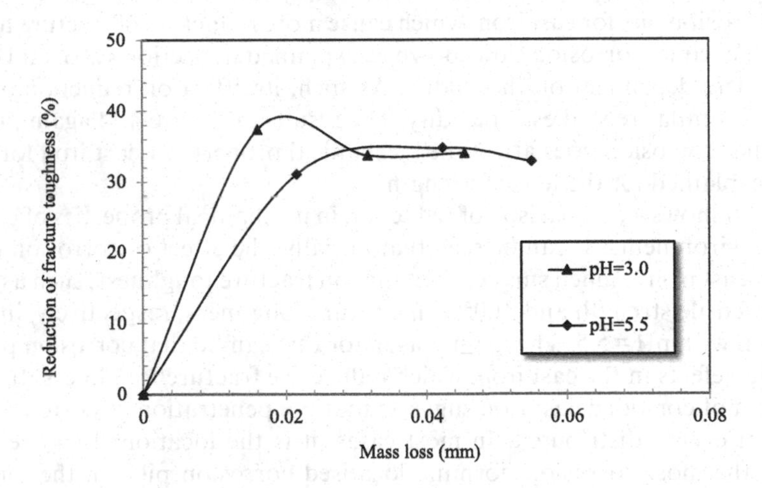

(b) With corrosion loss mm

Figure 4.14 Reduction of fracture toughness of cast iron with (a) immersion time and (b) mass loss.

Table 4.6 Comparison of Reduction of Mechanical Properties (%)

Time Period (Day)	*pH*	*Tensile Strength*	*Fracture Toughness*
90	3.0	8.45	37.37
	5.5	7.22	31.05
270	3.0	11.64	33.99
	5.5	9.89	32.79

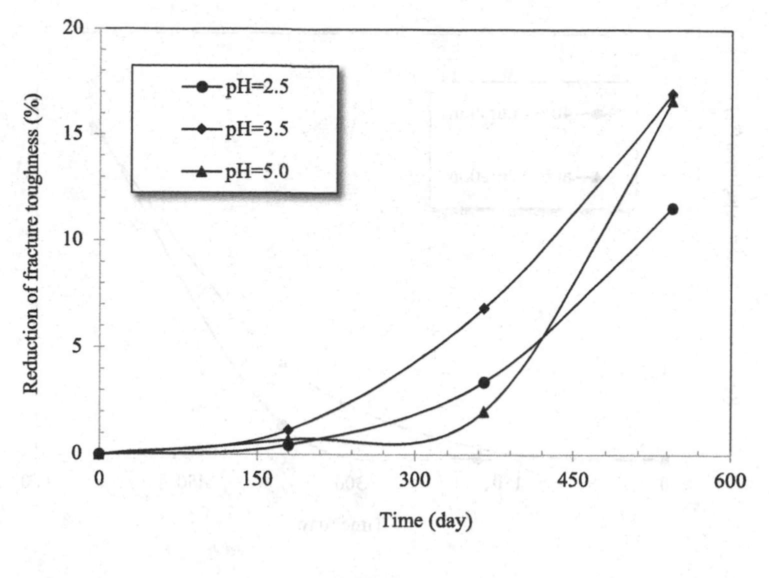

(a) High saturation

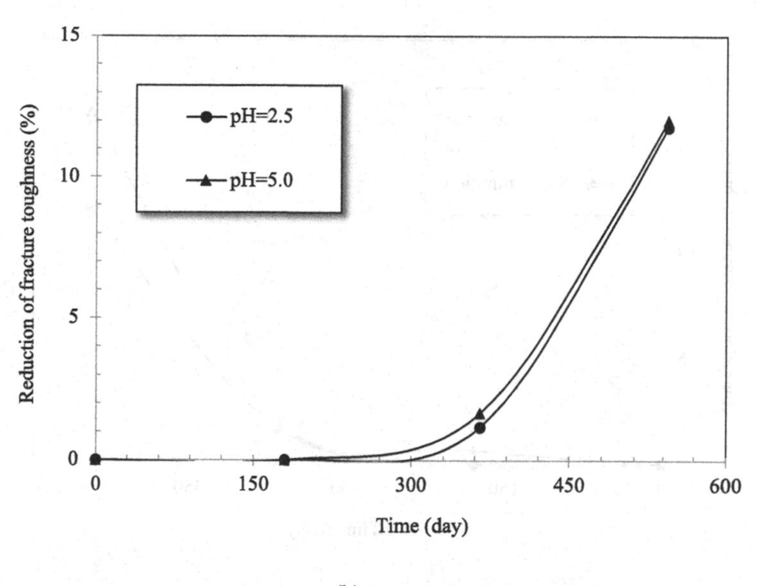

(b) Low saturation

Figure 4.15 Reduction of fracture toughness in soil with (a) high saturation (80%), (b) low saturation (40%), (c) high acidity (pH = 2.5) and (d) low acidity (pH = 5).

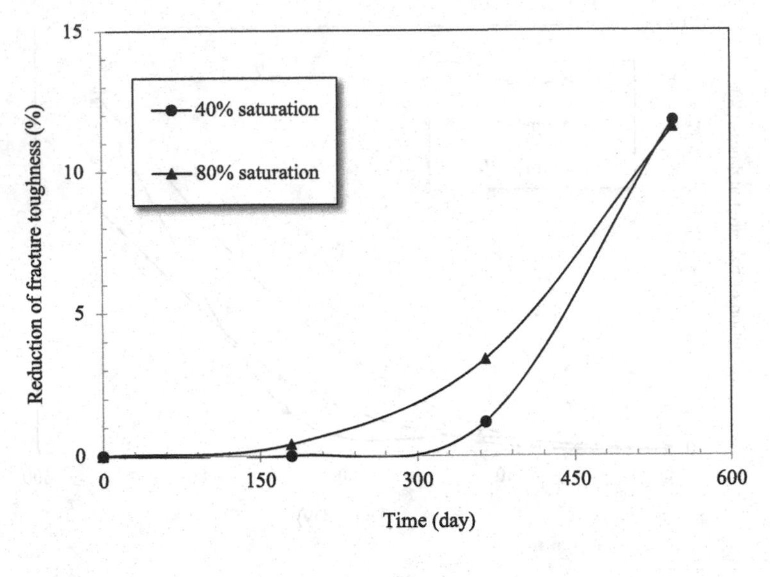

(c) High acidity

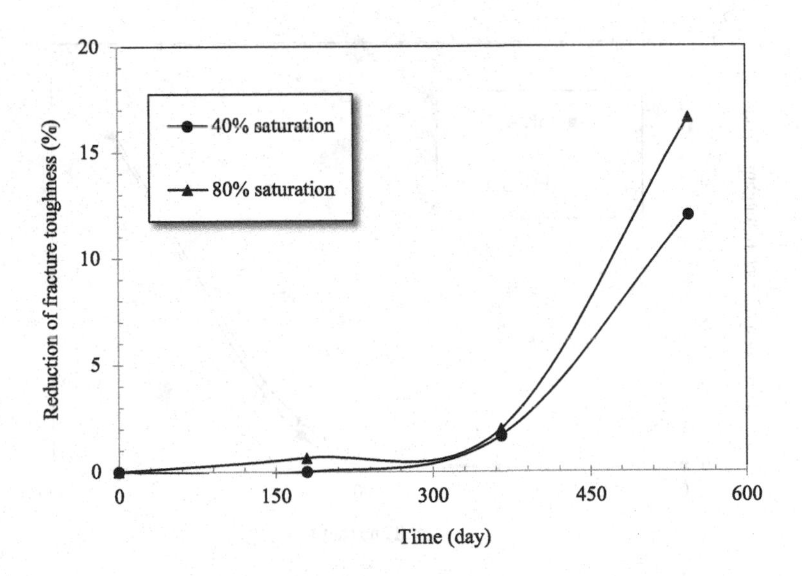

(d) Low acidity

Figure 4.15 (Continued) Reduction of fracture toughness in soil with (a) high saturation (80%), (b) low saturation (40%), (c) high acidity (pH = 2.5) and (d) low acidity (pH = 5).

(Wasim et al. 2020a). A comparison of the variation of the fracture toughness of corroded cast iron in various soil conditions is also made.

From Figure 4.15a it can be seen that at the initial stage of corrosion, almost no reduction occurs in fracture toughness. The original value of fracture toughness of the uncorroded cast iron is 24.63 MPa·m$^{1/2}$. After 180 days of corrosion in the soil, the fracture toughness only decreases slightly by less than 1% (24.47 MPa·m$^{1/2}$) for all acidic conditions of soil with 80% saturation. After one year (365 days) of corrosion in the soil, however, the reduction of fracture toughness varies with the condition of soil, from 3.37% (23.80 MPa·m$^{1/2}$) in high acidic soil (pH = 2.5) to 6.82% (22.95 MPa·m$^{1/2}$) in moderate acidic soil (pH = 3.5) and to 1.99% (24.14 MPa·m$^{1/2}$) in low acidic soil (pH = 5.0), respectively. After one and half years (545 days) of corrosion, the reduction of fracture toughness continues to vary with the condition of soil, from 11.5% (21.79 MPa·m$^{1/2}$) in high acidic soil (pH = 2.5) to 17% (20.46 MPa·m$^{1/2}$) in moderate acidic soil (pH = 3.5) and to as large as 16.5% (20.54 MPa·m$^{1/2}$) in low acidic soil (pH = 5.0), respectively. One feature of interest in Figure 4.15 is that the reduction of fracture toughness is not proportional to acidity at all times. This is different to the reduction of tensile strength where higher acidity induces more reduction. The reason can be that the reduction of fracture toughness is more related to local damage by corrosion, such as cracking, in particular at the notches of the specimens.

From Figure 4.15b it can be seen that in the soil with low saturation, the effect of acidity on reduction of fracture toughness is insignificant. The fracture toughness does not start to decrease until after one year of corrosion with about 1.7% (24.22 MPa·m$^{1/2}$) reduction. After one and a half year (545 days) of corrosion, the fracture toughness reduces by 12% (21.67 MPa·m$^{1/2}$) from that of the uncorroded cast iron. From Figure 4.15c, it can be seen that in the high acidic soil, the effect of saturation of soil is not significant except for some variation in the mid-process of corrosion. The important information from this figure is that in the end, the reduction of fracture toughness tends to be the same in high acidic soils with both high and low saturations. Figure 4.16d shows a similar trend of reduction of fracture toughness without much influence of soil saturation in the low acidic soil except that on the contrary to that for high acidity, high saturation impacts more on reduction of fracture toughness than low one in the longer term of corrosion, given the same low acidity. The reduction of fracture toughness is 16.5% (20.54 MPa·m$^{1/2}$) and 12% (21.67 MPa·m$^{1/2}$) in soils with both high (80%) and low (40%) saturation, respectively, for the given acidity of pH = 5.0. These results suggest that soil saturation affects the fracture toughness of cast iron more than acidity in the longer term. It appears that the localised reaction causes more reduction in fracture toughness.

The overall results in Figure 4.15 suggest that the fracture toughness of corroded cast iron reduces more in soil with high saturation than with high acidity. The reduction of fracture toughness may be due to the development of corrosion pits on the surface of cast iron, which facilitates crack growth and propagation. This means that the localised reaction of corrosion contributes more to the reduction of fracture toughness of cast iron. Other fundamental reasons, such as changes in elemental composition and microstructure, will be discussed in Section 4.5 below.

Results in Figure 4.15 have practical implications in the design and assessment of corrosion and its effect on the mechanical properties of cast iron buried in soil. Figure 4.15 shows that soil with low acidity (pH = 5.0) and high saturation (80%) has the largest impact on reduction of fracture toughness (16.6%). It also shows that, in addition to sectional or material loss due to corrosion as widely recognised in the field of cast iron corrosion in soil, corrosion can reduce the resistance of cast iron to fracture. Results in Figure 4.15 in fact explain very well the sudden bursts of corroded cast iron pipes during operation, resulting in catastrophic consequences to the public.

The reduction of fracture toughness of cast iron taken from section specimens is shown in Figure 4.16. As described in Section 4.2.1, the dimension of the section specimens is 120 (length) × 120 (diameter) and 11 (thickness) mm. The section is buried in the soil with various acidity values (pH = 2.5, 3.5 and 5.0) and 80% saturation that are the same as those for coupon specimens so as for direct comparison. After corrosion, a sample is cut from the section, and specimens for fracture toughness tests are made following the same standard as for coupon specimens. From Figure 4.16a it can be seen that the fracture toughness of cast iron buried in soil with all conditions reduces. The general trend is the same as that with coupon specimens. What is of interest here is that the largest reduction of fracture toughness is in soil with high acidity (pH = 2.5) and high saturation (80%), which is different from the results from coupon tests. The reason could be that relatively there are more localised reactions with coupons than with section specimens due to perhaps larger surface areas of the latter. As reasoned above, localised reaction has more impact on reduction of fracture toughness. After one year of corrosion, the maximum reduction of fracture toughness is 12.9%, which is larger than that from tests on coupons with the same condition.

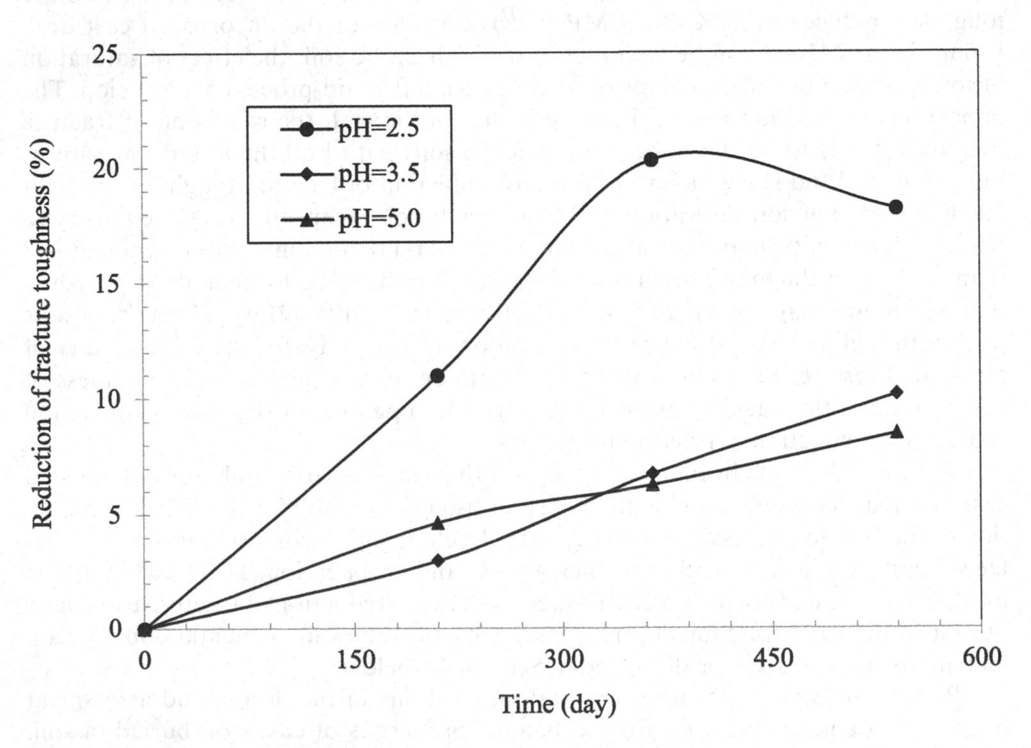

Figure 4.16 Reduction of fracture toughness from section specimen.

Information on corrosion of cast iron in natural soil environment and its effect on fracture toughness reduction is not widely available. Again, opportunity was seized to obtain pipes exhumed from service on which corrosion and its effect on the reduction of fracture toughness are studied. Table 4.4 shows the pipes exhumed from service after a number of years of operation. A piece of plate is cut form these pipes and cleaned in accordance with a standard. After corrosion measurement, specimens are made from the plate for fracture toughness tests. Figure 4.17 shows the results of tested fracture toughness where each point represents an average of 2–3 test results. It can be seen from the figure that after service of 37 years, the fracture toughness degrades by about 8% due to corrosion. After 79 years of service, the fracture toughness degrades by 16.8%, which means in the second half of designed service life, the degradation of fracture toughness is doubled. Compared with the reduction of fracture toughness in soil solutions, it can be inferred that the maximum reduction of fracture toughness due to corrosion is about 16%–18%. Overall, the results of fracture toughness reduction in different corrosive environments are quite similar. From a phenomenological approach as it is taken in this book, the focus is on observation of end result, i.e., degradation of fracture toughness from a mechanistic perspective. This information can be more useful to practical assessment and prediction of failures of corrosion-affected cast iron and cast iron structures (pipes).

For practical application of test data on corrosion-induced reduction of fracture toughness of cast iron buried in natural soil, it is desirable to develop an empirical relationship between measurable parameters of corrosion, such as corrosion loss, and the reduction of fracture toughness. From Figure 4.17, the empirical relations can be derived from regression analysis as follows:

$$\Delta K_c = 7.18C \tag{4.3}$$

where ΔK_c is the relative reduction of fracture toughness (as defined by Equation (3.7)) in %, and C is the corrosion loss in mm. The coefficient of determination of the regression is $R^2 = 0.96$, which is acceptable.

It may be of interest to see how different mechanical properties of cast iron degrade in the same exposure environment. Such comparison is made and shown in Figure 4.18. It can be seen from the figure that in the same exposure environment, the fracture toughness of cast iron degrades the most, followed by tensile strength and then modulus of rupture. These results are not a surprise because corrosion of cast iron is associated more with pits and cracking which impact more on fracture toughness than tensile strength and modulus of rupture. The least degraded is the modulus of rupture maybe because it is a section or ring with some effect of structural system.

It needs to be noted again that the test results presented in the paper are one step towards establishing understanding and knowledge on corrosion effect on the mechanical properties of ferrous metals, such as cast iron in this section. The significance of these results lies more in its trend qualitatively than absolute values quantitatively. It is acknowledged that more tests are necessary to produce a larger pool of data for sensible quantitative analysis, based on which to develop theories and models for corrosion- induced deterioration of mechanical properties of ferrous metals. Obviously, this is ongoing work.

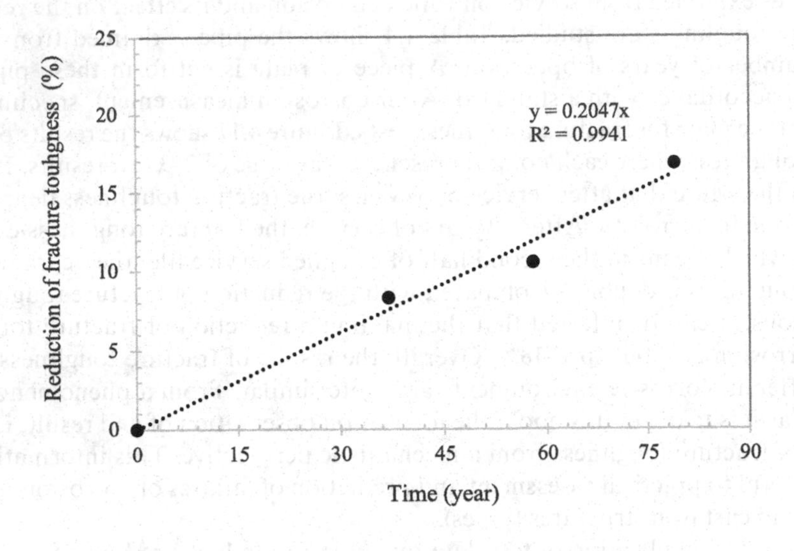

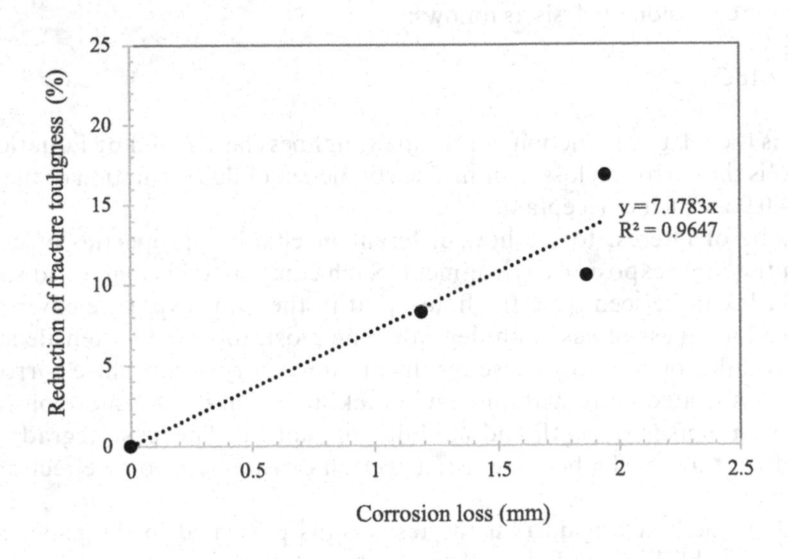

Figure 4.17 Reduction of fracture toughness of cast iron in natural soil with (a) time and (b) corrosion loss.

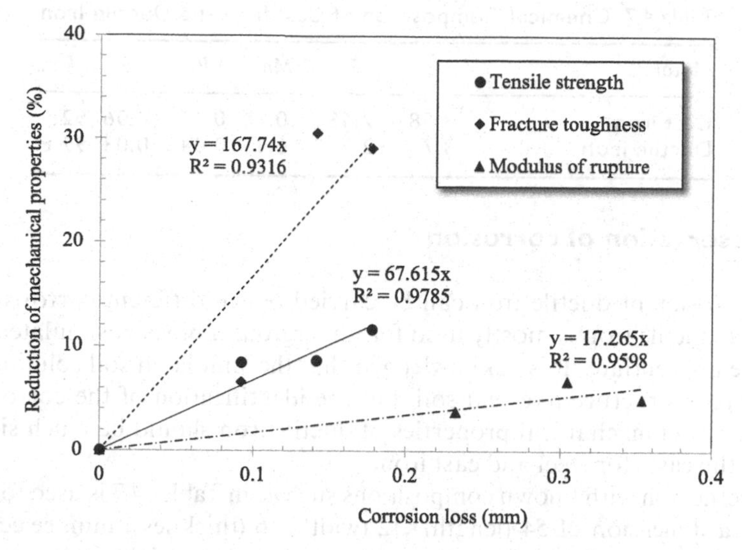

Figure 4.18 Comparison of degradation of different mechanical properties of cast iron in the same corrosive environment.

4.4 DEGRADATION OF MECHANICAL PROPERTIES OF DUCTILE IRON

Ductile iron as a building material came into use in the 1950s primarily to overcome the ductility issue of cast iron, which is very brittle. The process of manufacturing ductile iron is different from that for cast iron in the way that magnesium is added, which causes the carbon in the metal melt to precipitate upon solidification in the form of graphite nodules within the ferritic alloy matrix. The mechanical properties of ductile iron, such as ductility, strength and fracture toughness, can be enhanced by heat treatment, which eliminates the brittle microconstituents produced during the casting process (Rajani and Kleiner 2001).

Information on corrosion of ductile iron is even lesser than that for cast iron. Ductile iron is mostly, if not exclusively, used for underground pipes. There are other applications of ductile but not in the capacity of structural material, i.e., loadbearing. Ductile is an improved version of cast iron. In fact, ductile iron is a type of cast iron. As discussed in Chapter 2, the main difference that differentiates cast iron from ductile iron is perhaps in the shape of carbon or more precisely the graphite in the ferritic matrix (Fe). The graphite in ductile iron has a nodular or spheroidal shape.

Table 4.7 shows an example of the chemical composition of cast iron and ductile iron. It can be seen that the five major chemical elements are more or less the same for both metals. The key difference in elemental composition is the content of sulphur (S) and phosphorus (P) which are smaller in ductile iron: the content of sulphur in ductile iron is about half that in cast iron, and the content of phosphorus is less than tenth that in cast iron. As shown in Table 2.1 of Section 2.1.2, both sulphur and phosphorus have a negative effect on the mechanical properties of ferrous metals. Less content of them can improve the mechanical properties of ductile iron.

Table 4.7 Chemical Composition of Cast Iron and Ductile Iron

Metal	*C*	*Si*	*Mn*	*P*	*S*	*Fe*
Cast iron	3.58	2.48	0.74	0.67	0.06	92.9
Ductile iron	3.7	2.7	0.4	0.04	0.03	92.8

4.4.1 Observation of corrosion

Tests on corrosion of ductile iron can be carried out in different corrosive environments. Since ductile iron is mostly used for underground pipes, a simulated soil solution is more appropriate. It is acknowledged that the simulated soil solution does not have a complex structure like real soil, but the identification of the corrosion effect on degradation of mechanical properties of ductile iron should be much simpler in a solution as the cases for steel and cast iron.

The ductile iron with known compositions such as in Tables 4.7 is used for test specimens with a dimension of 54 (length) × 12 (width) × 6 (thickness) mm, selected on the basis of the thickness of ductile iron pipes commercially available. The configuration of the specimens should comply with a national standard for mechanical tests, such as ASTM E 1820 (2013d) on three-point bending (SENB) specimen requirements for fracture toughness tests. For the purpose of comparison with results from steel and cast iron, only one surface of the specimens is exposed in the soil solution, and all other surfaces are coated with rust guard epoxy and then wrapped with plastic tape in a similar fashion to the cast iron specimens to simulate external corrosion of buried ductile iron pipes.

Since the main purpose for corrosion tests on ductile iron is to compare with that of cast iron and steel, only one level of acidity for simulated soil solution is selected. During the immersion tests, the temperature and humidity of the environment are kept constant, similar to corrosion tests on the cast iron. In addition, as with the cast iron specimens in solution, three duplicates of ductile iron specimens are tested for intended measurement at the designated times. At the designated times, the specimens are taken out of the immersion, and corrosion measurements are taken, and then the specimens are tested for fracture toughness.

Figure 4.19 presents the results of corrosion rate of ductile iron in the simulated soil solution with pH = 2.5. It can be seen that in the first 180 days of immersion, the corrosion rate increases to 0.1 mm/year. Compared with that of cast iron, the corrosion rate reduces by about 15% (from 0.118 mm/year), which is considerable. What is more interesting is that in the next 185 years, the corrosion rate only increases about 30% to 0.13 mm/year. The trend of corrosion rate in the figure is similar to the conceptual model described in Section 2.2.2 for the first two stages. That is, the corrosion increases rapidly in the beginning and then flattens to the peak and decreases eventually although this is not shown in the figure due to lack of sufficient test data.

4.4.2 Reduction of fracture toughness

The reduction of tensile strength of ductile iron is quite similar to that of cast iron. Thus, it is not repeated in this section. But the reduction of fractur toughness of ductile

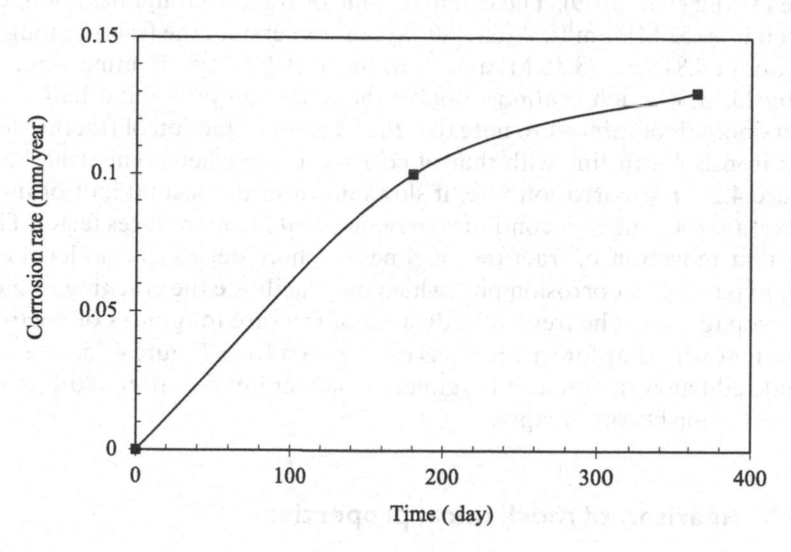

Figure 4.19 Corrosion rate of ductile iron in simulated soil solution.

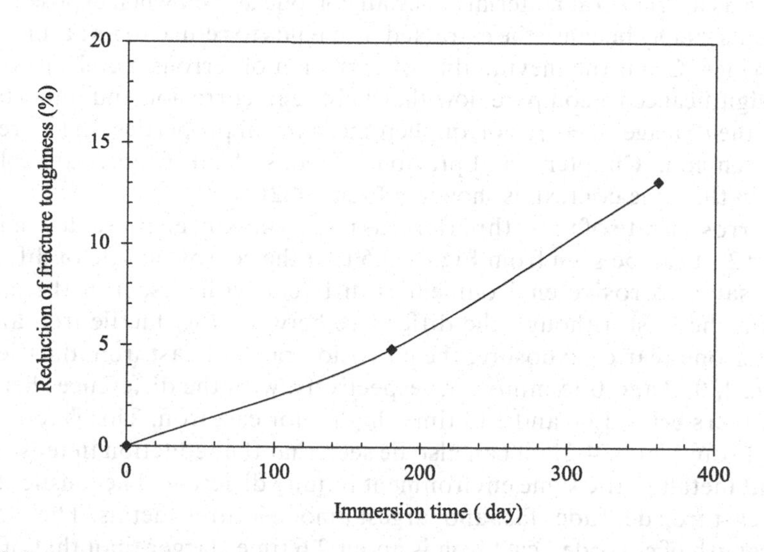

Figure 4.20 Reduction of fracture toughness of ductile iron in simulated soil solution.

iron is different to that of cast iron. As for cast iron, ductile iron specimens are taken out of the immersion at designated times, and after corrosion measurement, they are made for fracture toughness tests with the same configuration and procedure as those for cast iron.

Figure 4.20 shows the results of fracture toughness tests for ductile specimens in the simulated soil solution with pH = 2.5 where each point represents the average of two measurements. It can be seen that the fracture toughness of ductile iron reduces with time (Wang et al. 2019). The original value of fracture toughness of uncorroded ductile iron is 40.52 $MPa{\cdot}m^{1/2}$. After 180 days of immersion, the fracture toughness reduces by about 4.84% to 38.56 $MPa{\cdot}m^{1/2}$. In the next 195 days of immersion, it further reduces by 13.13%, which is almost double the reduction in the first half duration of the immersion. It is of interest to note that the trend of reduction of fracture toughness of ductile iron is not in line with that of corrosion rate when comparing Figure 4.19 with Figure 4.20. For corrosion rate, it slows down in the second half of immersion, but for fracture toughness, it continues to reduce and in fact reduces faster. This again indicates that reduction of fracture toughness is more dependent on local corrosion damages, in particular corrosion pits, which may facilitate the cracking initiation and eventual propagation. The trend of reduction of fracture toughness of ductile is however consistent with that for cast iron, as can be seen from Figure 4.15, where it shows that initial reduction of fracture toughness is moderate, but after 180 days of corrosion, the reduction becomes rapid.

4.4.3 Comparison of mechanical properties

Steel, cast iron and ductile iron are collectively referred to as ferrous metals in this book. They are all used as building and machinery materials with perhaps steel most widely used as a structural material. They all corrode as well when exposed to various environments simply because the extracted iron tends to return to its natural state, i.e., the iron oxides. Given the inevitability of corrosion of ferrous metals, it would be of practical significance to compare how they behave in corrosion and importantly how they resist the damage of corrosion on their mechanical properties. In this regard, test results presented in Chapter 3 and previous sections of this chapter are collated and presented in the same context as shown in Figure 4.21.

The corrosion rate of these three ferrous metals have been presented in Figure 2.5 of Chapter 2. It can be seen from Figure 2.5 that the corrosion rate of different metals in the same corrosive environment is different with cast iron the largest and ductile iron the least although the difference between the ductile iron and steel is small. After one year of exposure, the corrosion rates of cast iron, ductile iron and steel are 0.32, 0.13 and 0.15 mm/year, respectively, with the differences between cast iron and others being 1.46 and 1.13 times higher for cast iron. This is too significant to ignore. From Figure 4.21a it can also be seen that the reduction in tensile strength of different metals in the same environment is quite different. The tensile strength of corroded cast iron degrades fast and largest amongst three metals. The reduction of tensile strength of corroded cast iron is about 2.6 times larger than that of corroded steel. This makes sense and is understandable since the cast iron contains graphite flakes which facilitates corrosion in boundary of grains and intergranular corrosion. In the process of corrosion, graphite flakes also promote cracking. Together with boundary and intergranular corrosion, the integrity of the metal is damaged, and the mechanical properties degrade. Relatively, corrosion of steel is more uniform than cast iron which may have some pits due to less inhomogeneity of cast iron. Pitting can also cause cracking. Thus, the reduction of tensile strength is the largest for cast iron. Similar results are also reported in published literature, such as Dean and Grab (1985).

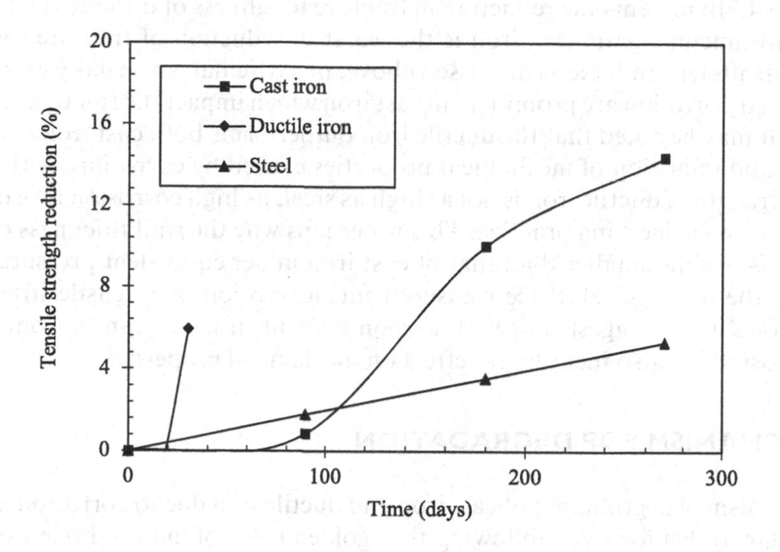

(a) Tensile strength

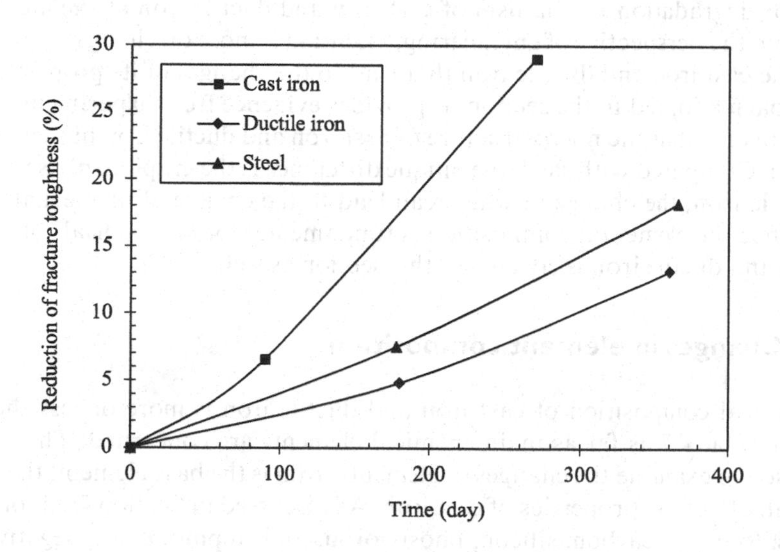

(b) Fracture toughness

Figure 4.21 Comparison of corrosion effect on the mechanical properties of ferrous metals: (a) tensile strength and (b) fracture toughness.

Figure 4.21b presents the reduction in fracture toughness of different metals in the same environment. Again, cast iron is the worst in reduction of fracture toughness. This is quite understandable as discussed above, in particular, since flaky graphite and intergranular corrosion are prominent in cast iron which impact the fracture toughness the most. It may be noted that the ductile iron outperforms both cast iron and steel in corrosion and reduction of mechanical properties caused by corrosion. Although the absolute strength of ductile iron is not as high as steel, its high corrosion resistance can be made use in engineering practice. This is perhaps why the wall thickness of ductile iron pipes is usually smaller than that of cast iron under equivalent pressure ratings. Moreover, the results of all three measurements (corrosion rate, tensile strength and fracture toughness) suggest that high carbon content in metal can not only lead to more corrosion but also incur larger effect on mechanical properties.

4.5 MECHANISM FOR DEGRADATION

The mechanism of degradation of cast iron and ductile iron due to corrosion is in principle similar to that for steel, following the "golden rule" of material science that the structure (atomic lattice) of a material determines the property of the material. Also, the current state of the art in corrosion science and engineering has not developed a theory for degradation mechanism of cast iron and ductile iron at the atomic lattice level. From the perspective of engineering, attention is more on changes of microstructure of the cast iron and ductile iron that lead to the changes of its properties. This is the approach adopted in the section. It provides evidence from experimental observation and results that the microstructure of cast iron and ductile iron has changed after corrosion. Compared with steel, the unique difference is the graphite phase of cast iron and ductile iron, the change of which can lead to the change of its mechanical properties. Since the elemental composition is fundamental for all materials, its change in cast iron and ductile iron is covered in this section as well.

4.5.1 Changes in element composition

The chemical composition of cast iron and ductile iron is more or less the same as shown in Table 4.7 as far as main chemical elements are concerned. Thus, only cast iron is used to examine the changes of elements. Iron is the base element, the change of which will affect the properties of cast iron. As discussed in Section 3.5.1, of five main elements, four, i.e., carbon, silicon, phosphorous and sulphur, could negatively affect the mechanical properties of cast iron. One element, i.e., manganese, may affect the mechanical properties of cast iron positively. There are other elements that penetrate cast iron during corrosion, for example, oxygen and chloride, which form the impurities in the cast iron after chemical reactions. Their presence in cast iron as impurities will negatively affect the mechanical properties of the cast iron.

As for the elemental analysis of steel, X-ray fluorescence (XRF) equipment, such as Bruker AXS S4 Pioneer XRF, is used to determine the element composition of cast iron. Selected results from the XRF tests for changes in contents of iron, oxygen, chloride, manganese and sulphur are presented in Figure 4.22 for cast iron buried in soil with two levels of acidity and 80% saturation (Wasim et al. 2019b). These environments

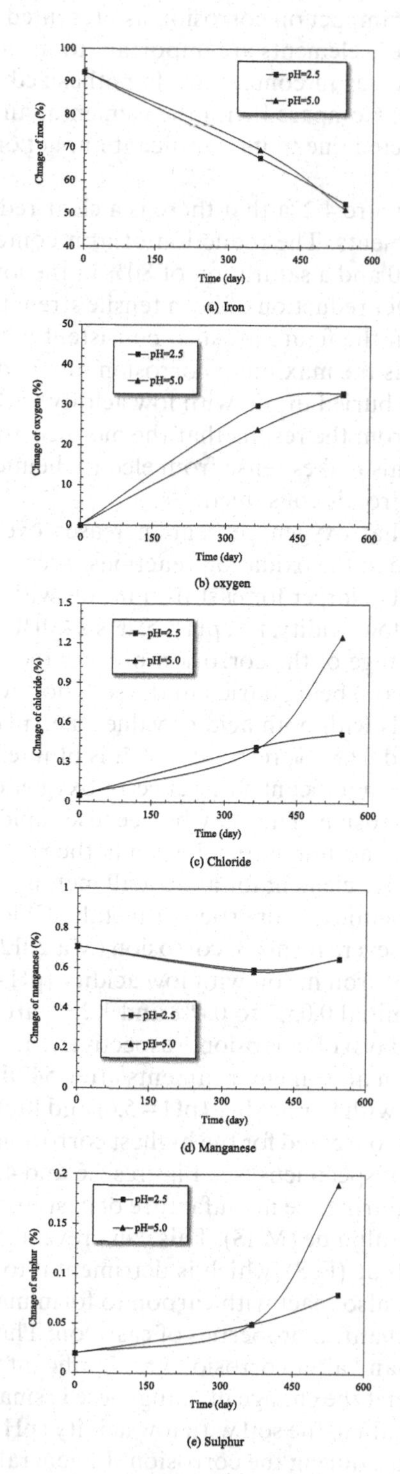

Figure 4.22 Changes of elements in cast iron in soil with 80% saturation: (a) iron, (b) oxygen, (c) chloride, (d) manganese and (e) sulphur.

are selected based on their impact on corrosion as presented in Section 4.2.3. Also, as discussed in Chapter 2, these elements are important to mechanical properties of cast iron, and hence, their changes in content are hypothesized to affect the mechanical properties of the cast iron. Compared with the elemental analysis for steel, one more element sulphur (S) is selected due to its significant impact on the mechanical properties of cast iron.

It can be seen from Figure 4.22a that there is a clear reduction of iron content in cast iron in both environments. The reduction of iron content is larger for cast iron buried in soil with pH = 5.0 and a saturation of 80% in the longer term. This is the environment that causes larger reduction of both tensile strength and fracture toughness of cast iron. The results in the figure are also consistent with the corrosion rate and pit depth measurements, as the maximum corrosion rate and depth are also observed for the cast iron specimen buried in soil with low acidity (pH = 5) and high saturation (80%). It can be inferred from the results that the more corrosion of cast iron leads to more reduction in iron. This makes sense from electrochemical reactions, the more of which there are, the more iron is consumed.

Figure 4.22b shows that oxygen content increases over time during corrosion, which is not a surprise due to the oxidation reactions. It can be seen that, initially, the increase of oxygen content is larger for cast iron in soil with high acidity (pH = 2.5) at 29.95%, whilst in soil with low acidity, i.e., pH = 5, it is 24.01%. This is because corrosion rate is higher at the first stage of the corrosion process requiring more oxygen. After one and a half years (545 days) being buried in the soil, however, the increase of oxygen content in cast iron in soils with both acidity values, i.e., pH = 2.5 and 5.0, is more or less the same at 33.16% and 32.89%, respectively. It is of interest to see that the acidity of soil does not make any significant difference in oxygen content in cast iron after one and a half years of corrosion. This may be because rapid corrosion occurs in high acidic soil at the beginning, accumulating oxygen in the rust layers of the cast iron.

Chloride is an aggressive element for almost all metals, and an increase of its content can damage the mechanical properties of metals. Chloride ingress has been reported as one of the most severe forms of corrosion (Ma 2012). Figure 4.22c shows that the chloride content in cast iron in soil with low acidity (pH = 5.0) and high saturation (80%) increases from an initial 0.05% to 0.4% and 1.25% after one year (365 days) and one and a half years (545 days) of corrosion, respectively. 1.25% is the highest chloride content in the specimens in all soil environments after 545 days of corrosion. This result confirms that the soil with low acidity (pH = 5.0) and high saturation (80%) is most prone for corrosion as also observed for the highest corrosion rate and pit depth of cast iron with different types of specimens (see Figures 4.6 and 4.8).

Manganese is added during the manufacture of cast iron to react with sulphur resulting in manganese (II) sulphide (MnS). This can prevent sulphur from reacting with iron to form ferrous sulphide (FeS), which is detrimental to mechanical properties of cast iron. Manganese can also react with carbon to form manganese carbide (Mn_3C), which can impact the mechanical properties of cast iron. Thus, the contents of Mn and S in cast iron both before and after corrosion are significant for its mechanical properties. Figure 4.22d shows that the change in manganese is small from initial 0.71% down to the maximum 0.59% again in the soil with low acidity (pH = 5.0) and high saturation (80%) although it fluctuates during the corrosion. In general, the decrease of Mn content would reduce the mechanical properties of cast iron.

Sulphur can be detrimental to mechanical properties of cast iron in particular to fracture toughness since it significantly reduces the ductility of metals. Figure 4.22e shows that the content of sulphur (S) in cast iron increases in all soil environments after one and a half years (545days) of corrosion. Again, the figure shows that the sulphur content in cast iron in soil with low acidity (pH = 5.0) and high saturation (80%) changes the most from the original 0.02% to the highest 0.19% after one and a half years of corrosion. Although the content in percentage is small (compared with other elements), the increase of sulphur is nearly ten times after one and a half years of being buried in the soil. As discussed in Chapter 2, the presence of sulphur may be detrimental to cast iron since it reacts with iron to form FeS, which reduces the bond strength amongst elements, directly affecting the integrity of the metal and hence the mechanical properties of cast iron.

Silicon is a vital element in all cast irons except white cast irons, as it allows the nucleation of graphite and minimises the cementite (FeC) content in cast iron. XRF analysis shows that silicon content fluctuates during corrosion in all burial environments, but the fluctuation is within 1%. There is no definitive trend of changes of silicon content in cast iron.

Changes of element contents in cast iron under accelerated corrosion environment can be verified by those in the natural environment. This can confirm at least qualitatively that corrosion does cause changes of elemental composition of cast iron. Figure 4.23 shows that in the corroded cast iron cut off from an exhumed pipe after 52years of service, the element content of base element iron (Fe) reduces and that of intruded elements oxygen (O) and chloride (Cl) increases. Also, the elemental content of manganese changes during the corrosion and eventually reduces although not very significantly. Compared with the changes of elemental contents in cast iron under an accelerated soil environment as shown in Figure 4.22, it is clear that the trend is similar or the same, which qualitatively verifies the results in accelerated soil environment. Overall, the results of element composition analysis of cast iron further confirm the fact that the elemental composition of cast iron changes due to corrosion. It suggests that corrosion-associated changes of material at a micro-scale level should be the main cause for the degradation of mechanical properties. The results from the current study are not only useful for the research community but also can help asset managers implement better management of corrosion-affected pipelines.

As noted in Section 3.5.1, the XRF analysis of elemental composition is on the exterior surface of corroded cast iron, which, depending on the depth of this surface, may not fully represent the composition of the bulk cast iron. The point is that at a certain depth into the bulk metal, the changes of elemental composition can affect the properties of the bulk cast iron. On the other hand, the mechanical properties, such as tensile strength and fracture toughness, do change after corrosion as phenomenologically observed in the tests and presented in the figures of Sections 4.3 and 4.4. It is believed that changes of element contents in cast iron or ferrous metals in general are one of the contributors for the degradation of mechanical properties due to corrosion.

Since ductile iron is a kind of cast iron, the changes of its element content are similar and hence are only briefly covered in the section. The changes of element content in five key elements of ductile iron immersed in soil solution with pH = 2.5 are summarised in Table 4.8 (Wasim 2018), where two 0 contents mean not detected.

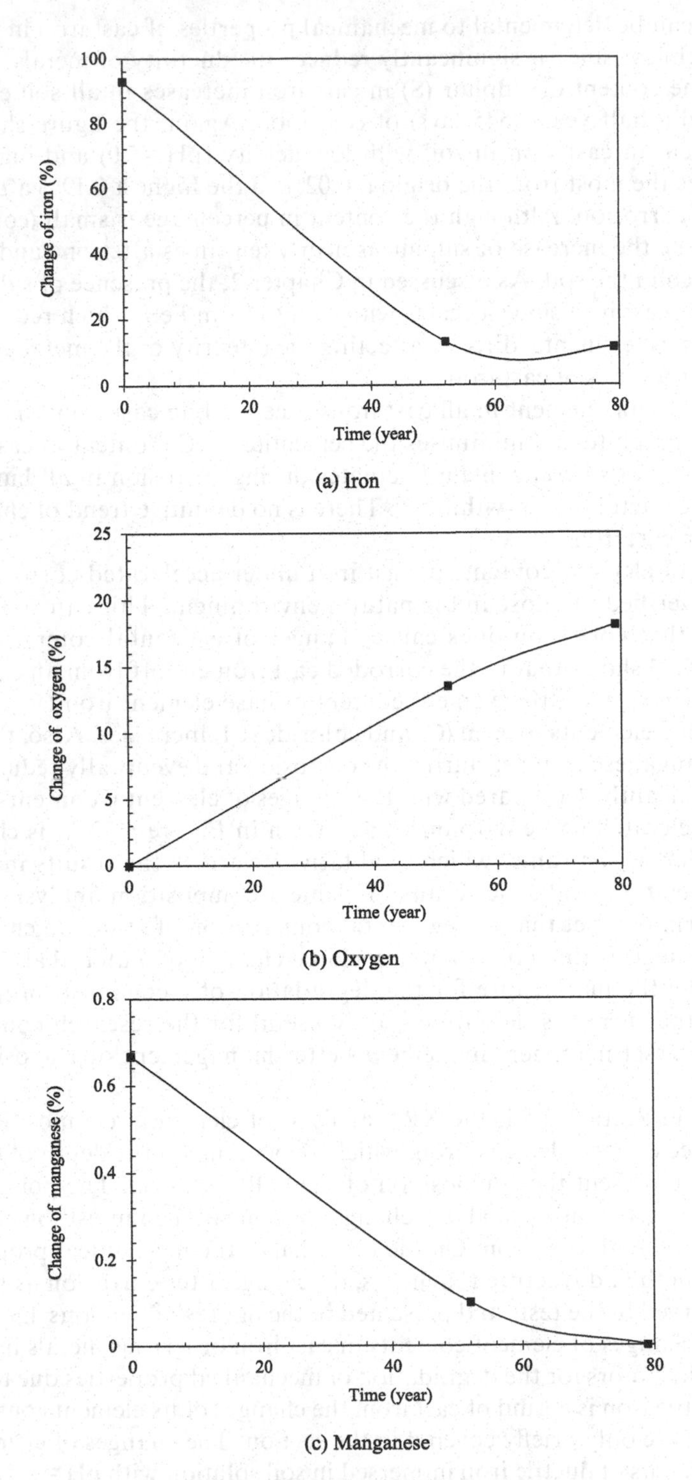

Figure 4.23 Changes of element content in cast iron in natural soil: (a) iron, (b) oxygen and (c) manganese.

Table 4.8 Changes of Element Contents in Ductile Iron (%)

Immersion (Days)	*Fe*	*O*	*Cl*	*S*	*P*
0	92.1	0	0	0.002	0.035
180	79.2	10.92	0.1	0.006	0.04
365	67.6	25.128	0.22	0.008	0.032

It can be seen from the table that the iron content reduces by 14% and 26% to be 79.2% and 67.6% from its original values of 92.1% after 180 and 365 days of immersion, respectively. This reduction is similar to that obtained for cast iron and mild steel immersed in the same soil solution. Similarly, like the other ferrous metals presented in the book, there is an increase in oxygen content from zero (not initially detected by the XRF) to 10.92% and 25.128% after 180 and 365 days of immersion in the soil solution, respectively. Furthermore, chloride content increases with time to 0.1% and 0.22%, respectively, after 180 and 365 days of immersion in the soil solution. The reduction of manganese content in ductile iron is similar to that of cast iron which is not very significant. Also, changes of element content in other elements, including sulphur, phosphorous and chromium are small as can be seen from the table.

In summary, corrosion-induced changes of elemental composition in cast iron and ductile iron can lead to the reduction of their mechanical properties. Cast iron and ductile iron consist of iron, carbon, silicon, phosphorus, sulphur, manganese and other elements with small quantities. These elements normally have different corrosion activations, and the active component (iron) is subject to preferential corrosion or dissolution. The selective corrosion of iron can reduce the ductility of the metal and eventually decrease the mechanical properties of the metal because less energy is absorbed during failure, such as rupture or fracture.

4.5.2 Changes in iron phase

Electron backscatter diffraction (EBSD) is used to quantify the changes in iron phases of corroded cast iron with time. The results of the corroded specimens from different environments are compared with each other and also with uncorroded cast iron to determine any phase changes due to corrosion over time. The EBSD measurements are performed with FEI Nova NanoSEM and Oxford Instruments with the Aztec software suite. A single EBSD mapping of a small scan area of 250 × 250 μm takes 12–16 hours for phase analysis.

Two main phases of iron, i.e., ferrite (α-Fe) and cementite (Fe_3C) plus inorganic crystalline iron oxide (Fe_2O_3), are selected for illustration of phase changes. Table 4.9 shows the changes of iron phases in cast iron after 1½ years (545 days) buried in the soil with various acidity and 80% saturation.

It can be seen from the table that, after one and a half years of corrosion, ferrite reduces sharply to 52.66%, and cementite reduces to 2.95%, whilst, not surprisingly, iron oxide increases to 28.23% in soil with low acidity (pH = 5.0) and high saturation (80%). This is the largest changes amongst all other soil conditions. As is noted, soil with

Table 4.9 Phase Changes in Cast Iron

Acidity in pH	*Saturation in %*	*Phase Analysis (%)*			
		α-Fe	*Fe_3C*	*Fe_2O_3*	*Other*
Uncorroded cast iron		81.18	3.56	0	15.26
5	80	52.66	2.95	28.23	16.16
3.5	80	72.32	2.42	13.63	11.63
2.5	80	64.50	2.95	21.23	11.32

low acidity (pH = 5.0) and high saturation (80%) also produces the largest corrosion rate and reduction of mechanical properties as presented in the previous sections. The phase analysis of these specimens is consistent with their elemental analysis, which suggests similar changes of iron (decreased) and oxygen (increased) contents. It may be noted in Table 4.9 that the effect of acidity on iron phase changes is mixed. It is not that the more acidic, the more changes. This phenomenon is also observed for corrosion rate and reduction of mechanical properties when cast iron is buried in soil with various acidity values as measured by pH. Since the acidity of pH = 5 is more realistic soil environment from a practical point of view, the results produced from this acidity can be useful and at least informative.

One of the distinct features of cast iron and ductile iron from steel is the phase of graphite. How corrosion changes the graphite can be analysed by SEM. Figure 4.24 shows the SEM images of cast iron specimens in various soil environments: (i) uncorroded, (ii) buried in soil with pH = 3.5 and 80% saturation and (iii) in natural soil for 37 years, i.e., samples taken from exhumed pipes. It can be seen form Figure 4.24b that the morphology of uncorroded cast iron is typically characterised by the presence of graphite flakes (i.e., the long black plates) in the matrix of iron. After corrosion, the morphology has changed with localised corrosion and graphitisation zones, which are the primary forms of deterioration for cast iron. The change of morphology indicates that, in addition to the corrosion-induced pits at the top surface of the corroded cast iron, corrosion penetrates the substrate of cast iron through the graphite flakes, causing degrading of their inherent properties. It is known that the presence of graphite flakes in cast iron can generate microcracks on the surface of corrosion pits (Conlin and Baker 1991). Since these microcracks can allow easy access of corrosion reactants from soil medium to the substrate, the corrosion process is accelerated (Wang et al. 2019). This type of microcrack can be seen in Figure 4.24b. As a result of combined element change, atmospheric oxidation of the corrosion products and intensified stress around pit front, the reduction of mechanical properties, in particular, the fracture toughness, of cast iron can be magnified.

For the sake of comparison, the microstructure photography of a specimen cut from an exhumed cast iron pipe after 37 years of service is presented in Figure 4.24c, where a thicker layer graphitisation zone can be seen. Figure 4.24c shows a resembling morphology of corroded cast iron in soil with 3.5 pH and 80% saturation (Figure 4.24b). This similarity in morphology suggests the effectiveness of corrosion tests in laboratory soil conditions. Overall, the results presented here are not only useful for researchers and practitioners but also can help asset managers implement better management of cast iron pipes through accurate corrosion assessments.

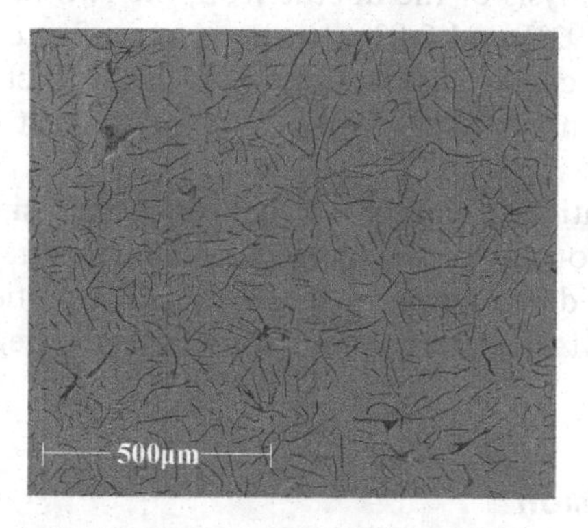

(a) Uncorroded

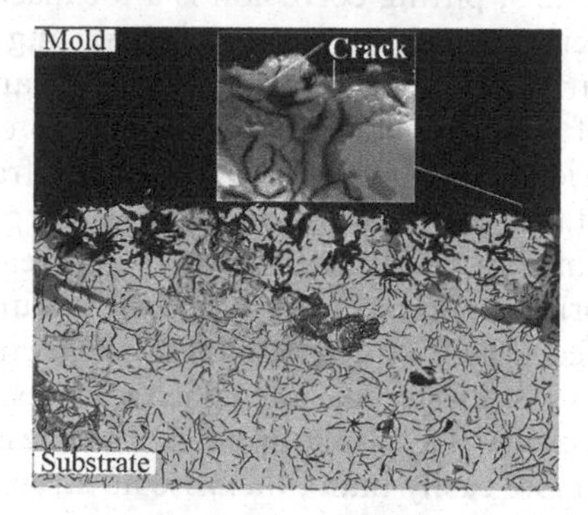

(b) Buried in acidic soil

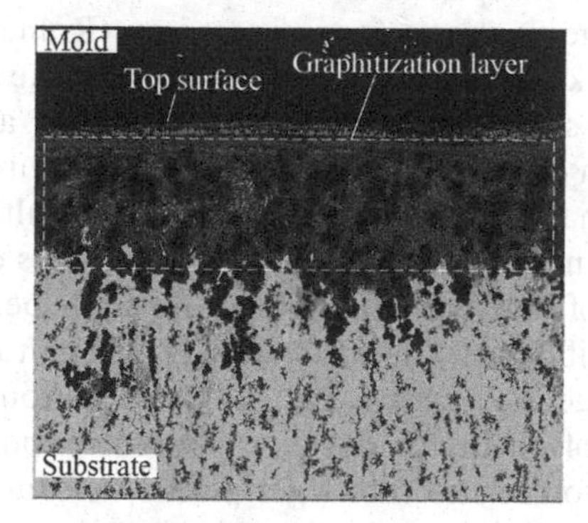

(c) In natural soil

Figure 4.24 SEM images of cast iron: (a) uncorroded, (b) buried in acidic soil and (c) in natural soil for 37 years.

As for the phase analysis of the ductile iron, the two main phases of ferrite and cementite reduces to 77.92% and 2.83%, respectively, after corrosion for 1 year. This is very similar to that of cast iron. The iron oxide in the ductile iron after one year of corrosion is 18.9%, which is slightly lower than that of cast iron due perhaps to less corrosion.

In summary, in addition to changes of chemical composition in cast iron and ductile iron, the phases of iron have also changed after corrosion. Together, these changes are the mechanisms for degradation of mechanical properties of corroded cast iron and ductile iron. At least they are the contributors to the degradation of mechanical properties.

4.5.3 Pitting corrosion

As presented in Section 2.2.3, pitting corrosion is a localised corrosion which is the most complex and damaging form of corrosion. Since pitting corrosion produces cavities or cracks on the surface of steel with certain depth and width, its damage to mechanical properties of ferrous metals in general and to cast iron in particular is significant. The most serious damage to these can be the fracture toughness due to the cavities or cracks that the pitting corrosion initiates and promotes. This is because fracture toughness is defined by crack extension. For this reason, this section focuses on the effect of pitting corrosion on the degradation of fracture toughness of cast iron.

Pitting corrosion affects the fracture toughness of cast iron in several ways. One of the most common ways is developing intensified stresses around sharp corrosion pits in cast iron when the external loads are applied. The stress concentration at the pit front can initiate cracks more easily than otherwise no stress concentration. As such, the fracture toughness of the cast iron can reduce in comparison with that without corrosion pits. Also, the atmospheric oxidation of the corrosion products left inside of corrosion pits can bulge on the surface of cast iron (Romanoff 1957). These bulges can further generate cracks, causing the reduction of fracture toughness of cast iron. When corrosion pits are sharp and narrow, they can play a direct role in initiating cracks in cast iron, which can more easily reduce the fracture toughness of cast iron. Furthermore, multiple sharp and narrow pits (such as multiple cracks) can interact with each other, causing multiaxial loading condition. This could further reduce the total fracture toughness of cast iron since the failure mode becomes non-Mode I. Even for a single corrosion pit, if it grows not transversely or longitudinally, the failure mode of fracture can also change. As a result, the total fracture toughness may be reduced.

However, the effect of pitting corrosion on fracture toughness of ductile iron is lesser than that of cast iron due to the larger plastic deformation, i.e., high ductility, exhibited by ductile iron. This is because the extensive plastic deformation around pitting corrosion in ductile iron will release the intensified stress before crack initiation or/and during the process of crack extension.

As the number of corrosion pits on the surface of cast iron increases, it may lead to a smoothly contoured region of thinning. Under this circumstance, although the stress concentration can be reduced, the effective cross-section area (thickness) of cast iron decreases. Since the effective cross-section of structures/specimens decreases, the loading condition may alter from plane strain to plane stress loading condition. This can change the fracture toughness of cast iron as well.

Another influence of pitting corrosion on the mechanical properties of cast iron and ductile iron is the atmospheric oxidation of corrosion products left inside of corrosion pits. As observed in the corrosion tests of cast iron, the presence of pitting corrosion can generate microcracks in front of corrosion pits. These microcracks can provide a path for corrosion reactants to trespass from the external environment (such as soil medium) to the substrate. This can accelerate the pitting corrosion particularly in cast iron with plenty of graphite that acts as an electrode in the corrosion cell, subsequently accelerating the reduction of fracture toughness of cast iron.

4.6 SUMMARY

Observations of corrosion of cast iron and ductile iron in different corrosive environments with focus on soil environment are presented in this chapter, which are simulated soil solution, real soil with various acidity and saturation values and natural soil. Different types of specimens and their effect on corrosion are also covered, namely coupon specimen and section specimen. Corrosion progress is expressed in time, in corrosion loss and importantly, in pit depth for cast iron. It is clear that corrosion is more active in soil with low acidity (pH = 5.0) and high saturation (80%). After corrosion tests, the effects of corrosion on tensile properties of cast iron are discussed first, including changes of stress-strain curves and degradation of tensile strength and modulus of rupture of corroded cast iron in different environments. It is clear that tensile strength and modulus of rupture of cast iron degrades during corrosion. Then, the test results on degradation of fracture toughness of both corroded cast iron and ductile iron in different corrosive environments are presented with focus on cast iron due to its significance in practical application. A comparison of corrosion-induced degradation of mechanical properties of different ferrous metals in the same corrosive environments shows that the fracture toughness of cast iron degrades most with steel the least. After that, the mechanisms for degradation of mechanical properties of corroded cast iron are explored from the perspective of elemental composition, iron phases and pitting corrosion. It is clear that corrosion changes the microstructure of cast iron and ductile iron and importantly causes pitting on the surface of the irons. As a result, the mechanical properties of cast iron and ductile iron degrade. The information presented in this chapter can be of great significance and interest to practitioners and researchers alike.

Chapter 5

Other corrosion damages

5.1 INTRODUCTION

In addition to the reduction of cross-sectional area of steel, which is well known, and reduction of its mechanical properties, which is presented in Chapters 3 and 4, corrosion also causes other damages. These damages can be exacerbated if the steel is under stress. As is known, corrosion takes place on steel when and perhaps only when it is exposed to some environments. Steel may not be exposed to the environment if it is not used in structures. Since the primary function of a structure is to carry loads which produce stresses in steel, it is reasonable to argue that all corrosion occurs simultaneously with stress. However, the stress effect on corrosion and subsequent damages to steel have not been well understood because information on these is not widely available. This chapter will present some.

Steel is made through continuous casting, in which variation in solidification speed affects the microstructure, resulting in different corrosion resistance of steel between the inner region, i.e., middle and the outer regions, i.e., edges. Corrosion due to intrinsic differences in the microstructure of steel is known as preferred corrosion in the steel manufacturing industry (Chilingar et al. 2013). Preferred corrosion is localised, non-uniform corrosion which causes stress concentration. The localised stress concentration initiates cracking in the steel, the scale of which can be such that the steel can split completely form the middle. This phenomenon is referred to as preferred corrosion-induced delamination in this chapter. Delamination of steel completely destroys the integrity of steel as a building and machinery material, as widely reported by, e.g., Beidokhti et al. (2009) and Pantazopoulos and Vazdirvanidis (2013). Although the preferred corrosion has been well known, its effect, i.e., delamination, is not. This knowledge can be gained through simulated corrosion tests and detailed microstructural analysis of the corroded steel as to be presented in this chapter. Previous corrosion tests for continuously cast steel only focused on the impact of manufacturing defects (e.g., central segregation, voids and cracks) on the corrosion process (Kajatani 2001). This represents only a small proportion of steel with poor manufacturing quality (Thomas 2001), which is diminishing due to increased quality control and advances in steel making technology.

Hydrogen embrittlement is another severe damage to steel when subjected to corrosion which reduces the tensile strength and ductility of the corroded steel. Hydrogen embrittlement occurs due to the accumulation of hydrogen at voids or defects in steel, which subsequently increases inner pressure. There are in general two test methods for investigating hydrogen embrittlement and importantly its effect on mechanical properties of steel. One is hydrogen charging in which hydrogen is charged into the steel physically. The other is hydrogen absorption in which hydrogen is absorbed in the steel during the corrosion process. In this chapter, the second method is used in observation of hydrogen concentration in steel and its effect on the mechanical properties of steel. Comparisons are also made to study the pros and cons of these two methods.

As in Chapters 3 and 4, a phenomenological approach is adopted in examining corrosion damages to steel in this chapter with focus on the mechanistic perspective. This approach can provide useful information, either qualitative or quantitative, on corrosion impacts on mechanical properties of corroded steel at least in a relative manner.

The results presented in this chapter can provide some information on issues that have no or accepted conclusion. For example, it is found in research (Chalaftris 2003) that mild steel, i.e., structural steel, is by and large safe for hydrogen embrittlement. There are even views (Hardie et al. 2006) that steel with yield strength less than 350 MPa is immune from hydrogen embrittlement. However, as presented in Chapter 2, hydrogen embrittlement occurs due to the accumulation of hydrogen at voids or defects, which subsequently leads to inner pressure increase. For mild steel with yield strength greater than 350 MPa, it was found (Eggum 2013, Djukic et al. 2016) that mechanical properties reduced due to hydrogen embrittlement once charged with hydrogen. However, no evidence, either laboratory or field data, has been provided to support the view that there is a threshold for yield strength of mild steel, such as 350 MPa, under which the steel is not affected by hydrogen embrittlement. This again makes the chapter interesting and useful.

5.2 STRESS EFFECT ON CORROSION

In most cases, corrosion and stress occur simultaneously since steel, or ferrous metals in general, is used in structures which are designed primarily to carry load, and hence, steel is under stress. It may be intuitive to think that stress may interact with corrosion, and their combined effect on mechanical properties of steel can be exacerbated. This section will explore this question.

There is a commonly held view that the applied stress, especially elastic stress, would not affect corrosion and hence the mechanical properties of corroded steel. A thorough review of corrosion science and mechanics, however, suggests that stress applied to steel would affect corrosion by reducing the corrosion resistance of corroded steel (Ren et al. 2012, Xu and Cheng 2012). This seems to be supported by the field survey of corroded steel (Li et al. 2018, Wang et al. 2018). There are three main mechanisms for the reduction of corrosion resistance: (i) stress can break down the protective passive oxide film and enhance the dissolution rate of iron (Gutman 1998); (ii) stress can increase the strain energy on the surface of steel which makes the corrosive solution easier to penetrate (Ren et al. 2012); and (iii) stress can increase the deformation at grain boundaries, leading to dislocations and slips amongst grains where corrosion is facilitated (Gutman 1998, Wang et al. 2014). These three mechanisms may not only act individually but also interact with each other. The interaction of stress and corrosion can not only affect the corrosion progress but also the microstructure and mechanical properties of corroded steel (Li 2018). The conflicting views on the stress effect on corrosion makes it more necessary to provide evidence on how stress affects the corrosion and subsequent changes in microstructure and mechanical properties of corroded steel. In particular, quantitative information on stress effect is very necessary, and a comparison between stressed and non-stressed steel during corrosion is much needed.

5.2.1 Observation of stress effect

To observe the effect of stress on corrosion, standard immersion tests in hydrochloric acid (HCl) solution (ASTM G31-72 2004b) can be carried out on steel specimens with everything the same except the stress, i.e., two sets of steel specimens are made identically: one with stress and the other without stress. After the corrosion tests, microstructural analysis of corroded steel is carried out according to ASTM E3-11 (2017a). Then mechanical tests of corroded steel are conducted according to ASTM E8/E8M-16a (2016). Details of specimen dimensions, samples for microstructural analysis and mechanical tests are the same as those presented in Chapter 3. For specimens with stress, the applied stress is 70% of the yield strength of the steel. This is selected so as to provide evidence that stress within the elastic range will or will not affect corrosion. A test rig is specifically designed as shown in Figure 5.1, where the specimen is pulled on a testing machine, and nuts are tightened to maintain the designated stress in the specimen (Li et al. 2019). Then, the test rig with the specimen under stress is placed in the HCl solution for corrosion tests. Of course, the frame that holds the specimen is either made of stainless steel or high strength steel coated with corrosion-resistant materials, such as epoxy, and wrapped with acid-resistant plastic to prevent the failure of the coating in acidic solution.

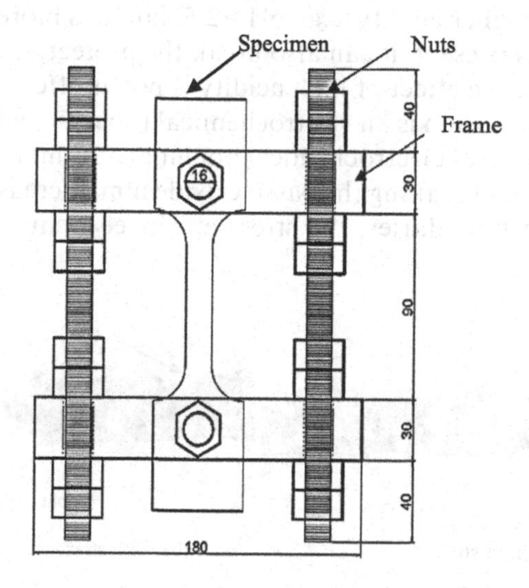

Figure 5.1 Details of test rig.

Three duplicate specimens are made for each test or measurement to ensure the repeatability and reproducibility of the test results and also to consider variation of corrosion and measurement, as well as their statistical analysis. For corrosion tests, three durations are selected to measure the corrosion progress over time which are 7, 14 and 28 days, respectively. Also, three levels of acidity, as measured by pH = 0, 2.5 and 5.0, are chosen to cover a wide range of possible acidic environments as described in Section 3.2.1. For example, the solution with pH = 5.0 is quite close to the natural corrosion of steel exposed to the environment containing a large amount of organic matter, e.g., steel buried in soil (Liu et al. 2014). At the end of each immersion period, specimens are taken out of the solution for measurement and testing. The main measurement of corrosion test is mass or thickness loss as described in Section 3.2.1 since it is uniform corrosion in immersion test. For microstructural analysis, the element composition, grain size and phase composition are examined.

Figure 5.2 shows the comparison of specimens with and without stress after 28 days of immersion in HCl solution with pH = 5. It can be seen that there are more rusts, more spalls and more pits on the stressed specimens. It is clear that there is an interaction of stress and corrosion, leading to increase of corrosion activities.

The corrosion progress as measured physically by corrosion thickness loss in mm for specimens with and without stress is shown in Figure 5.3, where each data point represents the average of three measurements. It can be seen that the corrosion loss of specimens with stress is 37% higher in a solution of pH = 5.0, 50% higher in a solution of pH = 2.5 and 44% higher in a solution of pH = 0 than that of specimens without stress, respectively, after 14 days of immersion in the same HCl solutions. Further, after 28 days of immersion, the corrosion loss of specimens with stress is 42% higher in the solution of pH = 5.0, 46% higher in the solution of pH = 2.5 and 47% higher in the solution of pH = 0 than that of specimens without stress. It is very clear from these figures that the stress has increased the corrosion activities significantly and consistently during corrosion progress for a range of corrosive environments. It may be noted that the effect of stress on corrosion is not proportionally increasing with acidity. The reason could be that higher acidity, e.g., pH > 2.5, initiates more corrosion than lower one, but when stress is present, it can also break the protective oxide film and initiate more corrosion. Thus, the effect of high acidity is not as effective as when there is no stress. As is known, corrosion is an electrochemical process, and the corrosion rate increases with the decrease of electrochemical potential (Gutman 1998, Revie and Uhlig 2008, Ren et al. 2012). By breaking the passive oxide film, increasing the surface energy and dislocating grain boundaries, the stress effectively reduces the electrochemical

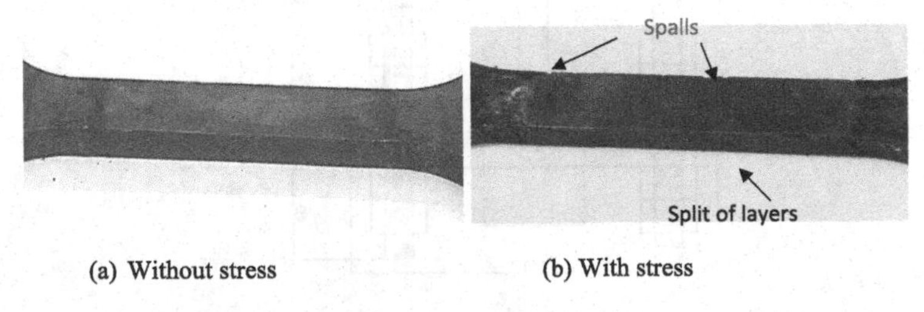

Figure 5.2 Photos of specimens after 28 days of immersion in HCl solution.

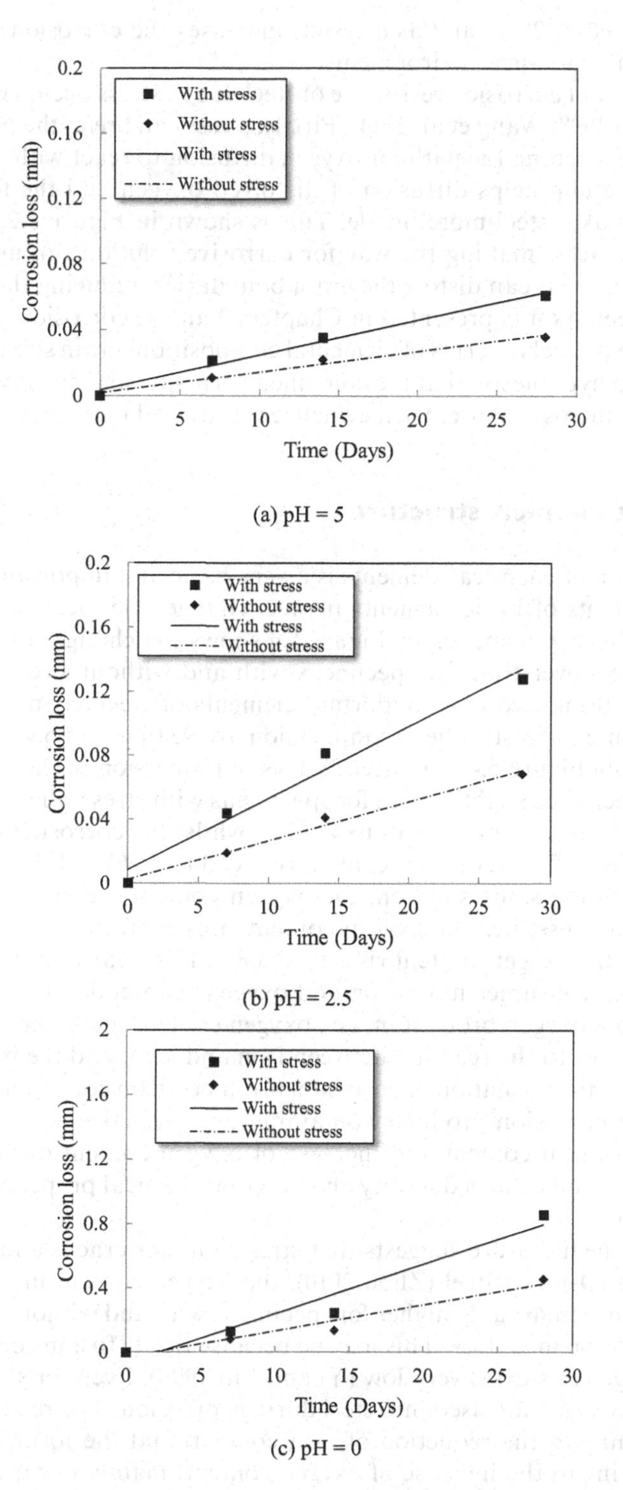

Figure 5.3 Comparison of corrosion progress in specimens with and without stress in various HCl solutions: (a) pH = 5, (b) pH = 2.5 and (c) pH = 0.

potential (Ren et al. 2012) and as a result, increases the corrosion rate as shown in Figure 5.3 for all corrosive environments.

The decrease of corrosion resistance of steel due to stress occurs with a few mechanisms (Gutman 1998; Wang et al. 2014). Firstly, stress can break the passive film on the surface of steel, opening the path for oxygen diffusion to react with iron. The rupture of passive oxide film helps diffusion of dissolved oxygen and the formation of rust layers, which makes steel more brittle. This is shown in Figure 5.2. Secondly, stress induces microcracks, making the way for corrosive solution through the steel (Ren et al. 2012). Thirdly, it can distort the grain boundaries, reducing the electrochemical potential of steel. As it is presented in Chapters 3 and 4, corrosion causes changes in microstructure of steel in terms of elemental composition, grain size and iron phase. It is reasonable to hypothesise that the combined corrosion and stress would cause more changes in the microstructure. Such evidence is provided in the next section.

5.2.2 Effect on microstructure

The composition of chemical elements is a very basic and important feature of steel. When the contents of basic elements in steel change, the mechanical properties of steel will be affected accordingly. Figure 5.4 shows the changes of element contents in corroded steel over time for specimens with and without stress immersed in the same solution, using two most important elements of steel as an example. The iron and oxygen contents in steel before immersion are 93.01% and 5.92%, respectively. It can be seen from Figure 5.4 that after 28 days of immersion in the same solution, the iron content decreases slightly more for specimens with stress when the corrosion loss is 67.93%, and iron content reduces to 48.63%, whilst the corrosion loss in specimens without stress is 50.86% with iron content reduced to 51.61%. Likewise, after 28 days of immersion in the same solution, the oxygen content increases slightly more for specimens with stress, i.e., the oxygen content rises to 47.27%, whilst for specimens without stress, the oxygen content rises to 46.08%. It is clear that stress induces more corrosion which consumes more iron and oxygen. Subsequently, iron content is reduced with more oxygen brought in, i.e., oxygen content increased. The reduction of iron content is due to the reaction between iron and acid, and the increase of oxygen content is due to the formation of corrosion products. Iron contributes to the ductility of steel, whilst corrosion products containing oxygen make steel brittle. Therefore, the reduction of iron content and increase of oxygen content during corrosion can lead to the reduction in steel ductility and other mechanical properties as to be shown in the next section.

Although the literature suggests that stress initiates cracks and facilitates diffusion of oxygen (O_2) into steel (Zhou 2010), the oxygen content in the corroded steel does not become remarkably higher for specimens with and without stress at the same degree of corrosion mass loss. This may be because the diffusion coefficient (1.2×10^{-2} $\mu m^2/s$) of oxygen in steel is very low (Yi and Lin 1990). Even for stressed steel, there is very limited oxygen diffused into steel during corrosion. The reaction between iron and acid (leading to the reduction of iron content) and the formation of corrosion products (leading to the increase of oxygen content) mainly occur at the surface for steel both with and without stress (Noor and Al-Moubaraki 2008). Therefore, the level

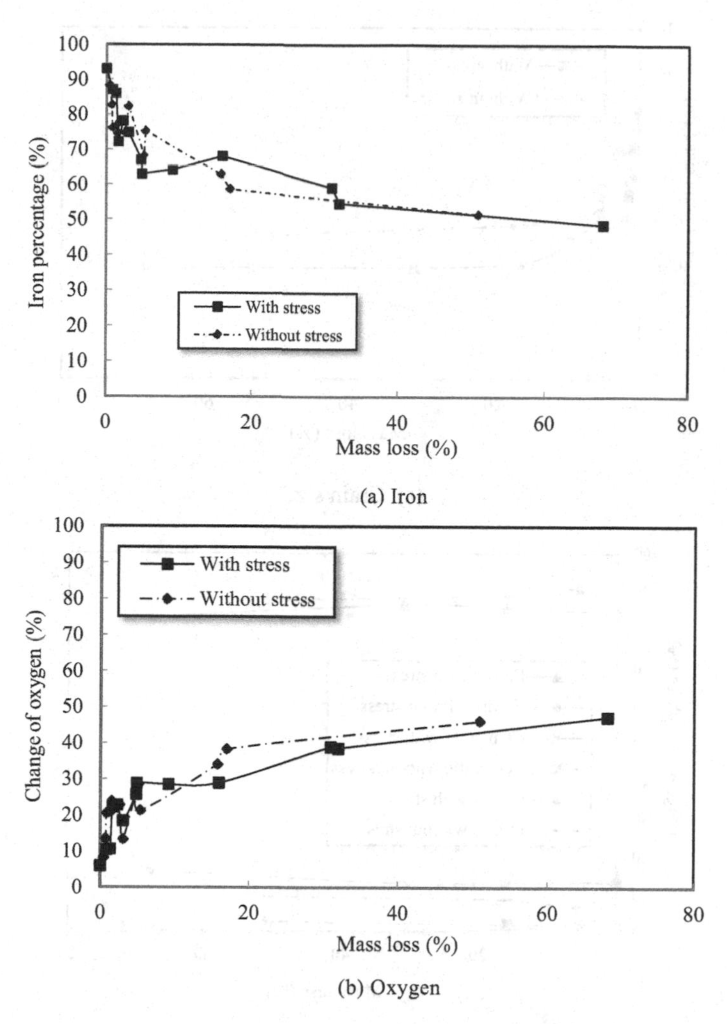

(a) Iron

(b) Oxygen

Figure 5.4 Change of element in specimens with and without stress: (a) iron and (b) oxygen.

of changes of iron and oxygen contents is, in general, not significant for steel with and without stress at the same corrosion degree.

As for the changes in grain size, test results show that, for specimens without stress, the grain size reduces by 29.94% in a solution with pH = 5.0, 40.07% in a solution with pH = 2.5 and 42.90% in a solution with pH = 0 after 28 days of immersion. For specimens with stress, however, the grain size reduces by 39.06% in a solution with pH = 5.0, 42.83% in a solution with pH = 2.5 and 58.73% in a solution with pH = 0, respectively, after 28 days of immersion. Figure 5.5a shows an example of the changes of grain size in steel immersed in HCl solutions. It can be seen that the average grain size of specimens without stress reduces from 12.18 to 6.52 μm when mass loss reaches 50.86%, whilst for specimens with stress, the grain size reduces to 5.01 μm when mass loss reaches 67.93%. For example, after 28 days of immersion, the reduction of grain

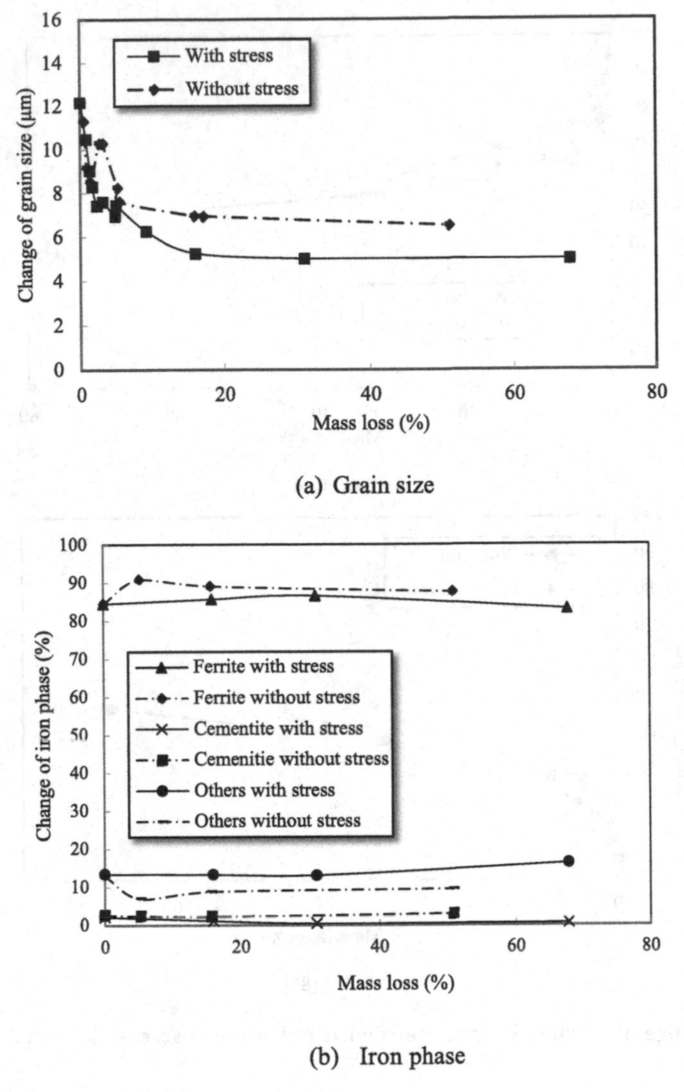

Figure 5.5 Change of microstructure in specimens with and without stress: (a) grain size and (b) iron phase.

size in steel with stress is 23% higher than that without stress. Also, at the same corrosion loss, the reduction of grain size in specimens with stress is about the same percentage (24%) higher than that in specimens without stress. The reduction of grain size is due to intergranular stress corrosion cracking (Li 2018).

Figure 5.5b presents an example of the changes of iron phase in steel immersed in the same solution but with difference in stress. It can be seen that the ferrite (α-Fe) content is around 85%, and cementite (Fe_3C) content is around 3% with the rest for other phases. With the corrosion progress, there are no significant and regular changes of iron phase composition in steel with and without stress during the corrosion. Although the corrosion resistance of cementite is larger than that of ferrite, corrosion mainly occurs at the boundaries of ferrite grains where cementite particles are located (Chisholm et al. 2016). As a result, cementite can easily be washed away by solutions. The composition of other phases, including graphite, austenitic and impurities within steel, can also be washed away by solutions since they are located at the boundaries of ferrite grains (Chisholm et al. 2016). Consequently, the level of reduction of ferrite, cementite and others is similar, and there are no significant changes of their contents in the ferric matrix. With the presence of stress, cementite is more likely to be fractured than ferrite (Umemoto et al. 2003). However, the intergranular stress corrosion cracking also contributes to the corrosion of ferrite grains, and other phases are washed away by solutions earlier (Arioka et al. 2006). As a result, the level of reduction of ferrite, cementite and others is still similar. It needs to be noted that information in Figure 5.5b is very little on changes of iron phase contents in steel with and without stress during corrosion. This can be another feature that makes the book unique.

As a summary of stress effect on the microstructure of steel, it is clear that changes are observed in all three microstructural features of steel, i.e., element composition, grain size and iron phase. Significant changes are observed in grain size with about 25% difference between the steel with stress and that without stress during corrosion. The changes of element composition and iron phase are not significant.

5.2.3 Effect on mechanical properties

From a mechanistic perspective, it is plausible that stress would affect the mechanical properties of steel during or after corrosion, in particular, fatigue strength and fracture toughness. To prove this by theories of mechanics can be very difficult. Thus, again, this section provides experimental evidence on stress effect on the mechanical properties of corroded steel, using tensile properties as an example since other mechanical properties are by and large related to tensile properties. If stress affects the tensile properties of corroded steel, it is most likely that it will affect other mechanical properties, e.g., fatigue and fracture toughness discovered in Chapters 3 and 4.

The stress-strain curve can represent the tensile property of steel comprehensively. Figure 5.6a shows such an example of stress-strain curve for specimens in the same HCl solution with only difference in stress. It can be seen that the reduction in ultimate strength and ductility is noticeable in specimens with and without stress after corrosion. The reduction of ultimate strength and ductility is caused by changes in microstructure under the combined corrosion and stress effects. Element composition, such as reduction of iron content caused by corrosion reactions, reduction in grain size caused by intergranular corrosion and the stress concentration at corrosion pits are the main factors that contribute to the changes of the stress-strain curve.

Figure 5.6b shows the reduction of yield strength against corrosion loss for specimens with and without stress in the same acidic solutions. It can be seen that there is a slight difference between steel with and without stress, although the pattern is very irregular. In fact, for specimens with stress, the yield strength increases with corrosion

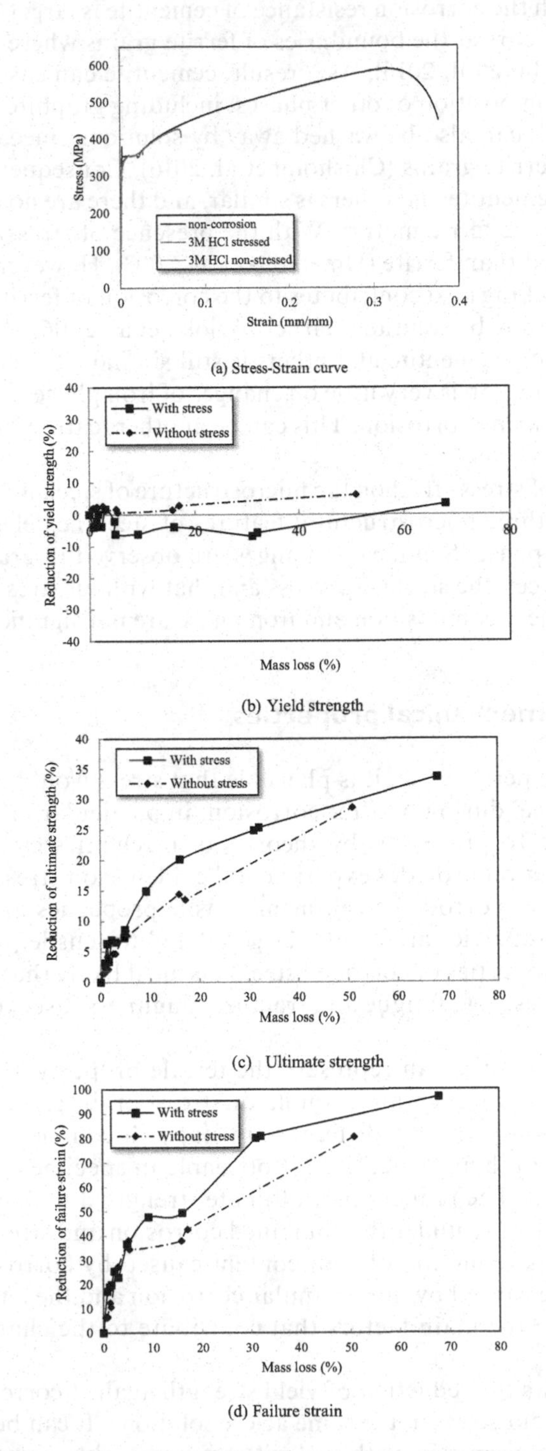

Figure 5.6 **Change of tensile properties: (a) stress-strain curve, (b) yield strength, (c) ultimate strength and (d) failure strain.**

initially for mass loss from 0% to 31% but eventually reduces. This can be that the pre-stress of 70% yield strength causes strain hardening due to some plastic deformation, adding energy to the steel (Li et al. 2019). Although the stress at 70% of yield strength is still in the elastic range, stress concentration at corrosion pits, defects and grain boundaries can lead to local plastic deformation and dislocation/slip amongst grain boundaries. Having said that, the net reduction of yield strength, i.e., from the increased yield strength to the final yield strength after corrosion, is about 9%. Compared with the maximum reduction of 6.2% for specimens without stress, the stress causes more reduction for corroded steel.

For ultimate strength, however, there is a clear difference between specimens with and without stress as shown in Figure 5.6c. It can be seen that the ultimate strength of specimens with stress reduces 1.17 times more than that of those without stress at the same degree of corrosion. It appears that changes of elemental composition and grain size play a more dominant role in this reduction than the strain hardening on the yield strength. Finally, Figure 5.6d shows that the failure strain for specimens with stress decreases more than that for specimens without stress. It may not be a surprise to see that the failure strain for specimens with stress decreases in a similar proportion (1.18 times) to that for specimens without stress at the same degree of corrosion, i.e., 40% of mass loss. This is mainly because the causes for failure strain reduction are the same as those for ultimate strength (Reive and Uhlig 2008).

To further demonstrate the stress effect on corrosion and mechanical properties of corroded steel, results for specimens without stress published by Garbatov et al. (2014) are obtained as shown in Table 5.1 and compared with those in Figure 5.6 using mass loss as a common measure of corrosion. It can be seen that the levels of reduction of yield strength, ultimate strength and failure strain in Figure 5.6 are larger than that for Garbatov et al. (2014), which are 1.65, 1.16 and 1.42 times larger, respectively, for specimens with 30% mass loss.

It is acknowledged that the results in Figure 5.6 are only for one stress level (70% of the yield strength). Ideally, more tests should have been carried out to study the effect of different stress levels on corrosion and mechanical properties of corroded steel since, in practice, corroded steel can be subjected to different stress levels during its lifetime. The point here is that stress does increase the corrosion and, together with corrosion, reduce the mechanical properties of corroded steel.

It may be of interest to use statistical tools, such as *t*-test (Devore 2012), to determine the significance of difference in two sets of test data; namely Data Set 1 from specimens with stress and Data Set 2 from specimens without stress. The objective is

Table 5.1 Stress Effect on Reduction of Mechanical Properties

Mass loss	*Stress*	*Yield Strength (%)*	*Ultimate Strength (%)*	*Failure Strain (%)*
20%	Without stress[a]	0.19	12.48	29.14
	With stress[b]	2.28	19.90	50.11
30%	Without stress	1.68	19.23	41.66
	With stress	2.77	22.38	59.10

[a] From Garbatov et al. (2014).
[b] From Figure 5.6.

Table 5.2 Statistical Analysis of Significance of Difference in Grain Size

Solution	*Immersion Days*	*t-Value*	*Accept or Reject Based on* $t_{0.05}$
pH = 5	7	2.19	Reject
	14	0.33	Accept
	28	2.12	Reject
pH = 2.5	7	5.21	Reject
	14	2.44	Reject
	28	0.87	Accept
pH = 0	7	5.33	Reject
	14	5.05	Reject
	28	5.54	Reject

to confirm whether stress affects corrosion or not. The null and alternative hypotheses can be stated as follows:

- Null hypothesis: Two sets of test data from specimens with and without stress at the same corrosion loss are not significantly different.
- Alternative hypothesis: One set of test data from specimens with stress is significantly different to the other set of test data from specimens without stress at the same corrosion loss.

For a confidence level of 95%, $t_{0.05} = 2.776$, the *t*-value for the test data from most of the cases is larger than $t_{0.05}$. Therefore, it is highly likely to reject the null hypothesis, which means that the test data produced from specimens with stress are significantly different from those from specimens without stress at the same corrosion loss. An example of *t*-test values for the significance of difference in grain size of specimens immersed in HCl solution is presented in Table 5.2. Details of this statistical *t*-test is beyond the scope of the book but can be referred to in other books on statistics, such as Devore (2012).

5.3 PREFERRED CORROSION

In the corrosion tests with acidic solutions, it is found that there is a split in the middle of the cross section or the thickness of the specimen after a certain degree of corrosion. Site inspections also observe this kind of split in the middle of cross section after steel corrodes to a certain extent. This phenomenon is caused by preferred corrosion which means that corrosion "concentrates" at the middle layer of the steel in terms of its thickness. Theoretically, corrosion due to intrinsic differences in the microstructure is defined as preferred corrosion. Preferred corrosion has been known in the steel industry, but there is limited knowledge – in particular, quantitative knowledge – on the causes of preferred corrosion (Revie and Uhlig 2008). Preferred corrosion can be catastrophic since it disintegrates the steel and, as a result, leads to structural collapse. Furthermore, there is little or no knowledge on the effect of preferred corrosion, that is, how preferred corrosion leads to steel delamination.

This section presents the causes for preferred corrosion, the factors affecting preferred corrosion, and how to prevent preferred corrosion. In order to identify the causes and influencing factors for preferred corrosion, three main microstructural characteristics are examined quantitatively at the middle and edge of the steel cross section, including grain size, iron phase and distribution of impurities. Corrosion also changes the composition of the elements in the steel, but the effect of elemental change on preferred corrosion is negligible (Marcus 2011, Li et al. 2018b). Thus, the effect of elemental composition on preferred corrosion is not covered in this section. Based on the analysis of causes and mechanisms of delamination, suggestions are put forward on how to prevent preferred corrosion from the steel making process and to the delamination of steel in structures. This section provides a quantitative understanding of the causes and effects of preferred corrosion on continuously cast steel, such as mild steel.

5.3.1 Causes of preferred corrosion

The cause of preferred corrosion is planted in the process of steel manufacturing. As it is known, steel is manufactured in most cases by continuous casting, also known as strand casting, (https://www.calmet.com, Vertnik and Sarler 2014), as schematically shown in Figure 5.7. Molten steel coming out of the blast furnace (see Figure 2.1) is conveyed into a scoop known as ladle. It is then poured into a large funnel, called tundish, which is located about 25–30 m above the ground level so that the casting process operates with the aid of gravity. The tundish is constantly supplied with molten steel to keep the process going. The molten metal is continuously passed through the mould at the same rate to match the solidifying casting. Further, the impurities and slag are filtered in tundish before they move into the mould. The entire mould is cooled with water that flows along the outer surface. Typically, steel casting solidifies along the walls of the casting and then gradually moves to the interior of the steel casting. The steel casting moves outside the mould with the help of different sets of rollers which also support the steel casting to minimise its bulging due to the ferro-static pressure. The sprays of water-air mist cool the surface of the strand between rollers to maintain the temperature of the steel strand until its molten core is solid. Whilst one set of rollers bend the steel cast, another set will straighten it. This helps to change the direction of flow of the steel slab from vertical to horizontal. The strand is then cut into slabs for structural use after the centre becomes completely solid (Thomas 2001).

Continuous casting is a method that was invented to streamline the production of steel with a view to reduce the cost of the production of steel. It also helps in the standardised production of steel, leading to better-quality steel products. Continuous casting eliminates some of the problems of traditional casting methods. For example, it eliminates piping and structural and chemical variations that are common problems of the ingot casting method. The solidification rate of the molten metal is also ten times faster than the solidification of the metal in the ingot casting method. Of many advantages over other processes of casting, the main advantage of continuous casting is its integration of several steps of casting, e.g., pouring of the molten liquid into casts, solidification and cast removal, into one congruent process which saves a considerable time of processing. However, this process embeds a severe defect in the steel, which is later discovered and known as preferred corrosion. More specifically, in

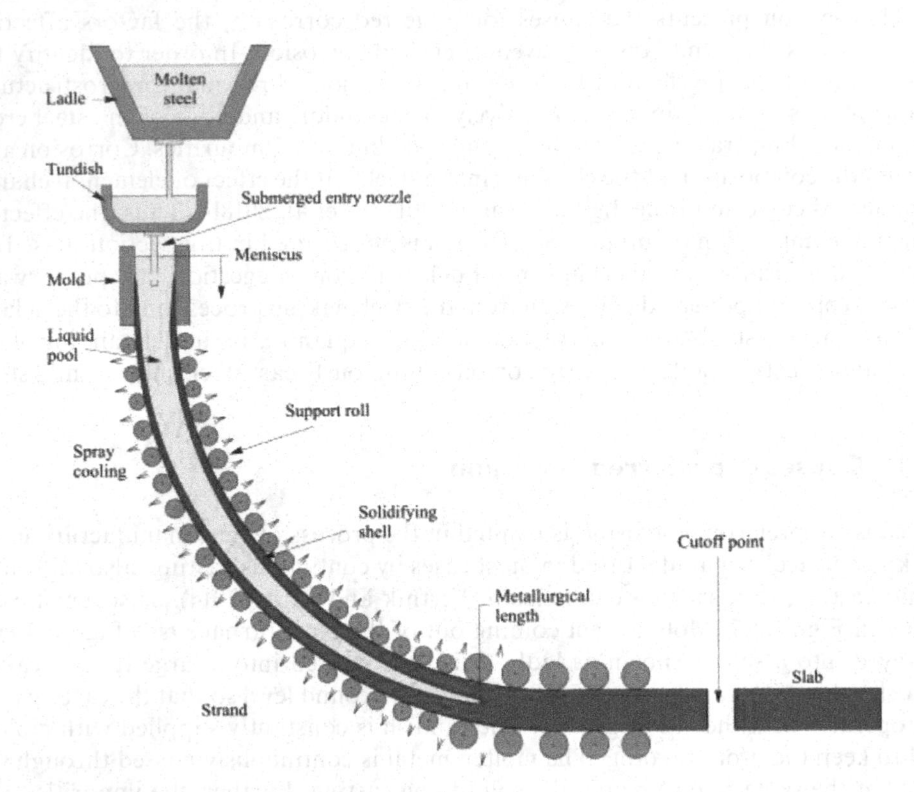

Figure 5.7 Continuous casting of steel.

continuous casting, molten steel at the edges solidifies faster than that in the middle of the thickness of steel (Thomas 2001). This difference in solidification speed affects the microstructure of steel and the impurities in it.

In continuous casting, depending on the thickness of steel product, the speed for molten steel to solidify can be different across the thickness of the steel. This difference in solidification speed affects the homogeneity of microstructure across the thickness of steel in three aspects (Shanmugam et al. 2007, Zhang and Thomas 2003): (i) grain size, (ii) iron phase composition and (iii) distribution of impurities. It is reported (Shanmugam et al. 2007) that the average grain size of steel products increases with decreasing speed of solidification. This seems to make sense intuitively. Steel contains two main phases in terms of its crystal structure, namely ferrite and pearlite. Ferrite is α-iron (α-Fe), which has a body-centred cubic structure, whilst pearlite is composed of 75% ferrite (α-Fe) and 25% cementite (Fe_3C). An increase in solidification speed can disperse pearlite into cementite and ferrite particles, which changes the iron phase composition of the steel. Molten steel contains other chemical compounds, such as oxides, which are collectively categorised as impurities. Zhang and Thomas (2003) suggested that dissolved impurities (mainly oxygen, aluminium and chromium) in molten steel precipitate when their concentration increases, which causes impurities to accumulate and reside in steel. The higher speed of solidification is the fewer impurities there are to accumulate and reside in the steel. For steel manufactured by continuous casting, the solidification speed for the middle of the steel strand is lower than that at the edges. As such, impurities will be concentrated and precipitated in the middle of the thickness of steel.

The microstructure of steel, as characterised by grain size, iron phase composition and distribution of impurities, affects its resistance to corrosion (Marcus 2011, Ralston and Birbilis 2010, Syugaev et al. 2008). Passive oxide films are formed before and during corrosion, which provide a protective layer for steel against corrosion (Marcus 2011). Smaller grain size and a larger proportion of cementite within the steel help to maintain the stability and adherence of these passive oxide films (Ralston and Birbilis 2010). Impurities in the steel accelerate the corrosion process by creating a concentration of local stress and galvanic reactions (Syugaev et al. 2008). For continuously cast steel, variation in solidification speed affects the microstructure, resulting in different corrosion resistance of steel in the inner region, i.e., middle and the outer regions, i.e., edges of the thickness of steel. Corrosion due to intrinsic differences in the microstructure is known as preferred corrosion in steel manufacturing industries (Chilingar et al. 2013).

Figure 5.8 shows the results of microstructure of intact steel with continuous casting, which confirm the difference in microstructure between the middle and edge across the thickness of steel. It can be seen from the figure that the grain size is smaller at the edge than in the middle, there is more cementite at edge than in the middle, and there are less impurities at edge than in the middle. All these differences make the middle of the steel more prone to corrosion. More quantitatively, the grain size at the edge is 6.36 μm and, in the middle, it is 12.18 μm, which is about twice the size at the edge. There is 1 impurity at the edge and 6 in the middle.

5.3.2 Factors affecting preferred corrosion

Of a number of factors that affect the preferred corrosion during the steel casting, the following three factors are considered as the main affecting factors.

5.3.2.1 Solidification speed

The speed of solidification during the continuous casting of steel can affect the microstructure of steel. With the process of continuous casting, the speed of solidification is different at each layer of the steel slab as shown in Figure 5.7. Steel surfaces solidify faster than substrate or inner layers, and the middle layers of the steel solidify slowest because the water spray cooling is performed on the steel surfaces. The difference in solidification speed makes the middle layer of steel have larger grain size, less cementite and more impurities than the edge. Large grain size and less cementite can affect the stability of passive oxide films and make them more susceptible to corrosion. Slow speed of solidification also incurs more impurities which can lead to galvanised corrosion. Thus, the middle part of steel has lower corrosion resistance than the edge, which leads to preferred corrosion. Preferred corrosion and its induced reduction in mechanical properties increase with solidification speed. The number of dislocations and manufacturing defects within the steel increase with solidification speed. This reduces steel's strength and ductility.

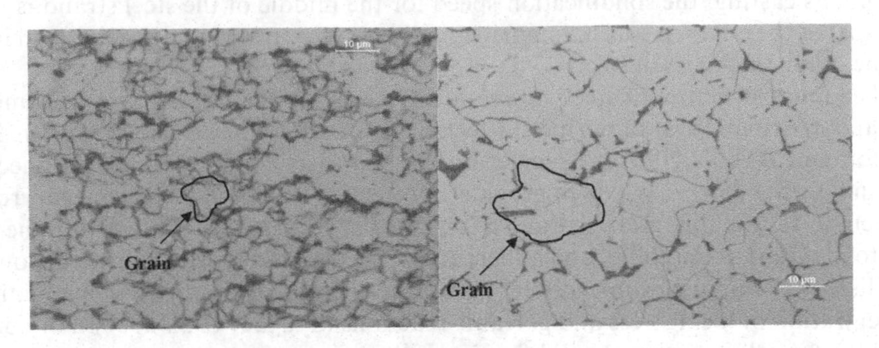

(a) Grain size

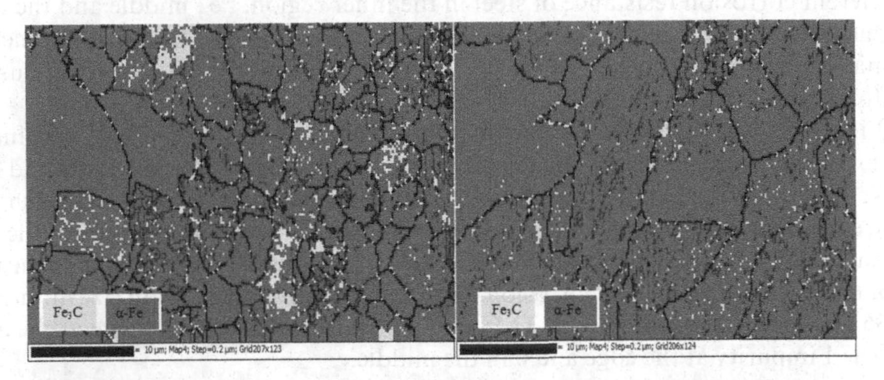

(b) Iron phase

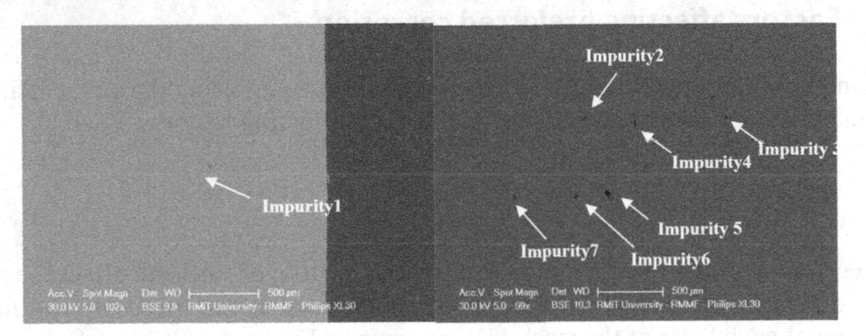

(c) Impurities

Figure 5.8 Difference in microstructure between the edge (left) and the middle (right): (a) grain size, (b) iron phase and (c) impurities (edge left and middle right).

5.3.2.2 *Elemental composition of steel*

As discussed in Chapter 2, the elemental composition of steel affects its corrosion which applies to preferred corrosion. Some elements in steel, such as chromium, molybdenum and vanadium, can contribute to the grain refinement during continuous casting. Passive oxide film stability can be increased if these alloy elements are added, which increases the corrosion resistance. An increase in chromium, molybdenum and vanadium can reduce the level of preferred corrosion and increase the yield strength and ultimate strength of both uncorroded and corroded steels. However, if too much vanadium is added within steel, it can reduce the ductility. On the other hand, carbon, aluminium, copper and sulphur can lead to localised corrosion of steel and reduce the corrosion resistance. An increase in carbon, aluminium, copper and sulphur content can contribute to preferred corrosion of steel. Although an increase in aluminium and copper can increase the yield strength and ultimate strength of uncorroded steel, it can reduce yield strength and the ultimate strength of corroded steel by causing preferred corrosion. However, a large amount of carbon can contribute to preferred corrosion and lead to a reduction in strength and ductility of corroded steel. A small amount of manganese can increase the corrosion resistance of steel by contributing to the grain refinement during steel manufacturing. However, if too much manganese is added, it can form compounds with oxygen and sulphur. These compounds can lead to a reduction in corrosion resistance by causing localised corrosion. Alloy elements mentioned above are likely to be concentrated in the middle layer of steel. Thus, they affect the difference in corrosion resistance between the middle and the edge and accordingly affect the preferred corrosion level.

5.3.2.3 *Temperature of steel casting*

An increase in temperature (above 187°C) can contribute to cementite formation within steel, especially in the middle layers of steel since they are the last part that cool down during manufacturing. A further increase in temperature (above 727°C) can contribute to austenite (face-cantered iron) formation within steel, especially in the middle. Both cementite and austenite can increase the corrosion resistance by increasing the stability of passive oxide film. The corrosion resistance in the middle of steel increases with the increase of casting temperature due to cementite and austenite formation. This reduces the preferred corrosion. An increase in casting temperature can also increase the yield strength and ultimate strength by forming these two phases. However, it can also lead to a reduction in steel ductility.

Due to these main factors during the continuous casting of steel, the microstructure of steel differs between the middle layers and edges, in particular, grain size, iron phase and impurities, as shown in Figure 5.8. With this knowledge, preferred corrosion could be prevented by homogenising microstructural characteristics and refining grain size in the middle layer of steel during its manufacturing process. The following suggestions could be beneficial to the steel industry in preventing preferred corrosion during the manufacturing of steel.

5.3.3 Prevention of preferred corrosion

Preferred corrosion occurs in the middle of the thickness of steel slab or cross section. It is caused by variations in grain size, phase composition and distribution of impurities between the middle and edge across the thickness of a steel cross section. Preferred corrosion can lead to steel delamination which disintegrates the steel and disable it as a structural material. In this sense, preferred corrosion-induced delamination of steel is the most severe form of steel deterioration since it destroys the integrity of steel. As discussed in Section 5.3.2, of many factors affecting or causing preferred corrosion, variation in the grain size of steel between the middle and edge across its thickness is the most dominant contributor. This is because the grain size in the middle is approximately twice the size of that at the edge (section Figure 5.8a). The root cause of preferred corrosion is, as presented in Section 5.3.2, the solidification speed of the casting process of molten steel. With this knowledge, preferred corrosion could be prevented by homogenising microstructural characteristics and refining grain size in the middle of the thickness during the manufacturing process of steel. The following suggestions could be beneficial in preventing preferred corrosion during the manufacturing of steel:

1. Preferred corrosion can be mitigated or prevented by conducting thermo-mechanical treatment on continuously cast steel products. The literature suggests that steel grains can be refined and homogenised using thermo-mechanical processing (Deb and Chaturvedi 1985, Junior et al. 2012). In this method, steel products are heated up to austenitisation temperatures (around 1,200°C) and immersed in a heat treatment fluid until they have been evenly heated. The heated steel is then strained under continuous cooling conditions and cooled down to room temperature. The steel grains are recrystallised during thermo-mechanical processing, which results in the grains being smaller and homogeneously distributed across the thickness of the steel.
2. Another method to reduce grain size and homogenise microstructural features is to perform equal-channel angular pressing (ECAP) on steel products (Shin et al. 2001, Valiev and Langdon 2006). This method presses steel samples repeatedly through a die with an L-shaped channel. The cross-sectional area of steel remains unchanged, and the steel is subjected to intense plastic straining during the process. By applying this exceptionally high strain, the steel grains are recrystallised, refined and homogenised. In addition, continuous ECAP procedures and plastic straining can be employed during the rolling process of continuous casting (Valiev and Langdon 2006). This results in a more homogeneous microstructure of the steel product, compared to steel made by conventional rolling, and accordingly prevents preferred corrosion from occurring.
3. The grain size of steel can also be refined by adding alloying elements (Maalekian 2007). For example, boron (B) accumulated at grain boundaries to form boron carbide $Fe_{23}(BC)_6$ during manufacturing prohibits the growth of the grain. In addition, niobium (Nb) and vanadium (V) precipitate during the rolling process of continuous casting and also hinder the growth of grains (Maalekian 2007). These alloying elements are likely to concentrate in the middle of thickness of steel during continuous casting (Thomas 2001). Therefore, it is recommended that B, Nb

and V could be added to steel during the manufacturing process to refine grain size in the middle of the steel and prevent preferred corrosion.

4. Eliminating the difference in composition of iron phase across the thickness of the cross section of steel also prevents preferred corrosion. The literature reports that this can be achieved by normalising continuously cast steel products (Digges et al. 1966, Shrestha et al. 2015). In this process, the steel is heated to austenitisation temperature, which makes ferrite and cementite grains recrystallised and transformed into austenite. After heating, the steel is held at the austenitisation temperature for a sufficient time to form a homogenous microstructure. Rapid cooling is then undertaken to decompose the austenite into ferrite, cementite and undefined oxides, resulting in them being uniformly distributed across the thickness of the steel. In addition, performing thermo-mechanical treatments on continuously cast steel products (as mentioned in point 1) can also help make the steel phase composition homogenous across its thickness (Junior et al. 2012).
5. It is also essential to control and try to eliminate impurities during the manufacturing process of steel. In continuous casting, molten steel goes through a solidification process where it flows out of the ladle and runs into mould through rollers. When the steel body is completely solidified, the steel is then cut into plates (see Figure 5.7). Deeper tundish increases the residence time of molten steel during manufacturing, which helps to remove impurities. Therefore, it is necessary to increase the depth of the tundish during the steel manufacturing process, in order to eliminate impurities and subsequent preferred corrosion.
6. Locations vulnerable to preferred corrosion in structural steel can be identified by examining the grain size distribution, iron phase composition and impurity distributions in the steel (as shown in Figure 5.8). Once these locations are identified, initiation and development of delamination can be predicted and prevented. For example, applying paint and catholic protection in these locations can typically prevent preferred corrosion from occurring.

It is acknowledged that the above measures for prevention of preferred corrosion and subsequent delamination can be time-consuming and costly. However, compared with the consequences of preferred corrosion-induced delamination, i.e., the loss of integrity of steel, collapse of steel structures and associated casualties and hazards, these measures are necessary and worthwhile for some structures.

5.4 CORROSION-INDUCED DELAMINATION

Preferred corrosion can destroy the integrity of steel as a building material through delamination as widely reported by, e.g., Beidokhti et al. (2009) and Pantazopoulos and Vazdirvanidis (2013). It is localised, non-uniform corrosion which causes stress concentration. The localised stress concentration initiates cracking in the steel, the scale of which can be such that the steel splits in the middle completely. This phenomenon is referred to as preferred corrosion-induced delamination in this section. There is little knowledge as to how preferred corrosion leads to steel delamination. There is almost no quantitative knowledge on preferred corrosion-induced delamination, such as where delamination starts, how deep and wide the delamination extends during corrosion and what the mechanism of delamination is. Furthermore, the initiation and propagation of corrosion-induced delamination can be affected by stress. When steel is under stress, such as in structures, the stress can accelerate corrosion as discussed in Section 5.2. It can also cause the microplastic deformation at grain boundaries and affect the distribution of residual stress across the thickness of steel cross section. Therefore, the effect of preferred corrosion can be more severe for steel with stress than that without stress. Since almost all steel is under stress, the combined effect of corrosion and stress needs to be considered.

The understanding of the cause and effect of preferred corrosion can be gained through simulated corrosion tests and detailed microstructural analysis of corroded steel. There is a clear need to conduct corrosion tests on continuously cast steel with new testing methodology and using specimens with no manufacturing defects to acquire quantitative knowledge on how the microstructure of steel affects preferred corrosion and how preferred corrosion leads to delamination of steel. Further, the combined effect of stress and corrosion on delamination is of interest.

5.4.1 Observation of delamination

There is little or no published literature to date on studies on corrosion-induced steel delamination. There is a clear need to provide evidence of how the microstructure of steel with no manufacturing defects affects preferred corrosion and how corrosion leads to delamination. Corrosion tests are conducted to observe the delamination of steel. Grade 250 mild steel is used for test specimen, which is continuously cast without any defects, as confirmed by the supplier. Dimensions of 90 × 14 × 6 mm are selected as test specimens to make their size suitable for the immersion tests and microstructural analysis. Following the standard sample preparation procedure, as described in previous chapters, three samples are cut from the specimen with the dimensions of 14 × 6 × 4 mm and prepared according to ASTM E3-11(2011) for microstructural analysis. The remaining parts of the specimens are used for general corrosion measurements (e.g., mass loss), the results of which have been presented in Chapter 3.

Corrosion tests are carried out in acidic solution following ASTM G31-72 (2004b). Same as in Chapter 3 for simulated corrosion tests, hydrochloric acid (HCl) solutions with three levels of acidity, as measured by its pH values, are selected in the corrosion simulation, namely pH = 0.0, 2.5 and 5.0. Although the primary reason to use a high acidic solution, e.g., pH = 0.0, is for accelerating corrosion, the range of acidity selected in immersion tests is not uncommon in the real-world situation as reasoned in Chapter 3. The duration of immersion is 7, 14 and 28 days after which specimens are taken out of the solution for various measurements. One specimen is in immersion until it is completely delaminated.

The main measurement for preferred corrosion-induced delamination is the opening width of the splitting layers of cross section and the penetrating depth in the steel body. An optical microscope (OM) and ImageJ (a software designed to edit and analyse images) are used to measure the width and depth of the delamination. After each period of immersion, three samples of 14 × 6 × 4 mm are cut from the specimens for microstructural analysis. Grain size is quantified for etched samples using an OM at 100× magnification for each immersion period. Three locations along the thickness of the cross section of the specimens are selected for measurement to examine the changes across the

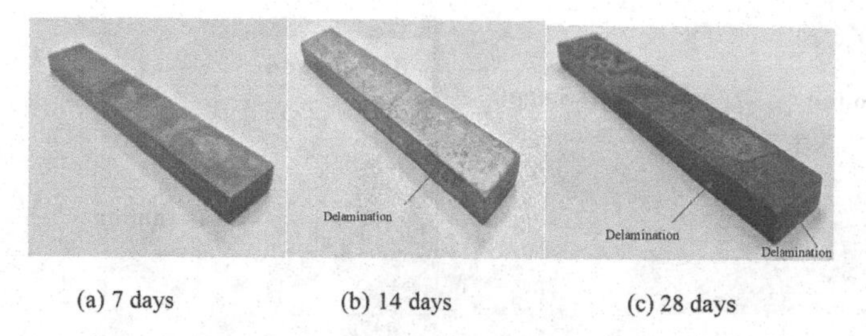

(a) 7 days (b) 14 days (c) 28 days

Figure 5.9 Photos of specimens after immersion in acidic solution with pH = 2.5 for (a) 7, (b) 14 and (c) 28 days.

thickness: (i) at the edge, (ii) at one quarter of thickness (1.5 mm from the edge) and (iii) in the middle. Iron phase analysis is carried out for both un-corroded and corroded samples at the same three locations using electron backscatter diffraction (EBSD) scanning. The scanning is conducted on three duplicate samples to determine the average phase composition. Since the pearlite comprises a two-phase structure composed of ferrite and cementite (Fe_3C), its proportion cannot be measured by EBSD directly. Therefore, the two phases selected for EBSD analysis are ferrite and cementite, which are known as the two phases that affect the corrosion behaviour as discussed in previous chapters.

Locations of impurities are determined using a scanning electron microscope (SEM) equipped with a backscatter electron (BSE) detector. The backscattering model of detection differentiates impurities from steel by showing them as different colours in the images (Pardo et al. 2008). To determine the changes in the number of impurities due to corrosion, BSE image analyses are carried out on both un-corroded and corroded samples and both with three duplicates at a magnification of 100×. The compositions of impurities are determined through energy-dispersive X-ray spectroscopy (EDS). The measurements are performed using Philips XL30 SEM at 30 kV voltages and 5.0 spot sizes.

Observation of specimens in acidic solutions indicates that there is no sign of split in specimens in solution with pH = 5.0 after 28 days. This can be understandable since corrosion in such solution is not severe enough to cause delamination of the steel. In other solutions, however, specimens in solution with pH = 0 show a sign of splitting after 7 days and specimens in solution with pH = 2.5 show a sign of splitting after 14 days. The photos of specimens after immersion in solution with pH = 2.5 are shown in Figure 5.9. It can be seen that after 14 days of immersion, delamination is clearly visible at the middle of the cross section of the specimen. Of more interest here is that after 28 days of corrosion, both sides of the specimens split and both at the middle of the cross section. This indicates that preferred corrosion occurs at the middle of steel cross section on all sides.

5.4.2 Quantification of delamination

The delamination is quantified in this section by opening width and penetrating depth of the splitting layers of the corroded steel. Figure 5.10 shows the OM images of the cross section (in yellow) of the specimens at 5× magnification after 14 and 28 days of corrosion. It can be seen that during corrosion, there is more mass loss in the middle

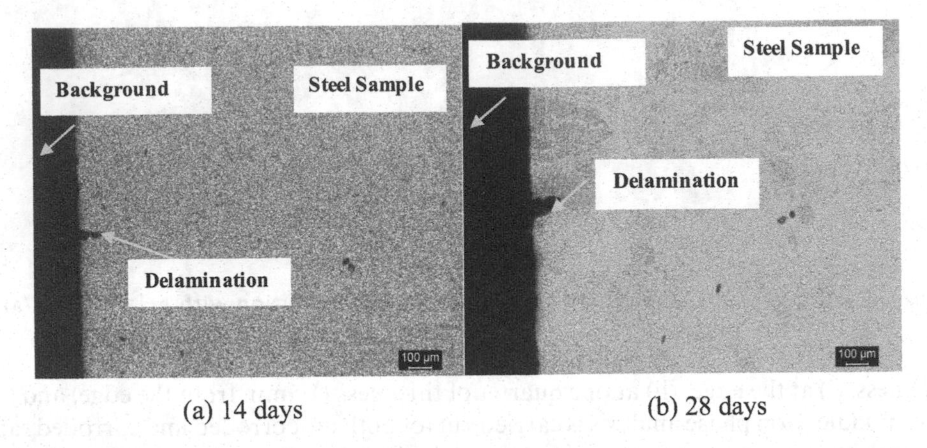

Figure 5.10 Mass loss in the middle region of the sample after immersion in solution with pH = 2.5 for (a) 14 and (b) 28 days.

region of the cross section of the steel sample than that at edge regions, forming a pit. This pit then propagates and extends inward as shown in Figure 5.10b. Eventually, further extension of the pit splits the steel sample.

Figure 5.11 summaries the width and depth of preferred corrosion-induced delamination of the steel sample for various periods of immersion in HCl solutions. It can be seen that, after 28 days of immersion, the width of delamination becomes 0.07 mm in the HCl solution with pH = 5, 0.13 mm in the HCl solution with pH = 2.5 and 0.76 mm in the HCl solution with pH = 0. Clearly, the degree of acidity plays a role in the delamination in a similar manner to corrosion. That is, the higher the acidity, the larger the deamination. For example, when the acidity increases from pH = 5 to pH = 2.5, the width of the delamination is almost doubled. When the acidity increases from pH = 2.5 to pH = 0, the width of the delamination is almost increased by six times.

Likewise, it can also be seen from Figure 5.11b that the depth of delamination of the steel sample increases from 0 up to 0.04 mm in the HCl solution with pH = 5, 0.13mm in the HCl solution with pH = 2.5 and 1.48 mm in the HCl solution with pH = 0 after 28 days of immersion. Again, when the acidity increases from pH = 5 to 2.5, the depth of the delamination is almost tripled. When the acidity increases from pH = 2.5 to 0, the depth of the delamination increases by more than ten times. It appears that the depth increases faster with the degree of acidity than the width of the delamination. The results in Figure 5.11 clearly show that the width and depth of delamination grow more rapidly under higher concentrations of HCl solution. Delamination is visible in the HCl solution with pH = 2.5 and very obvious in the HCl solution with pH = 0 after 28 days of immersion. Together with other figures, these results can be used to assess the severity of delamination and subsequently to predict the failure of corroded steel due to delamination.

5.4.3 Mechanism for delamination

From the manufacturing perspective, the speed of solidification of molten steel appears to be one of the main causes of preferred corrosion which leads to eventual delamination.

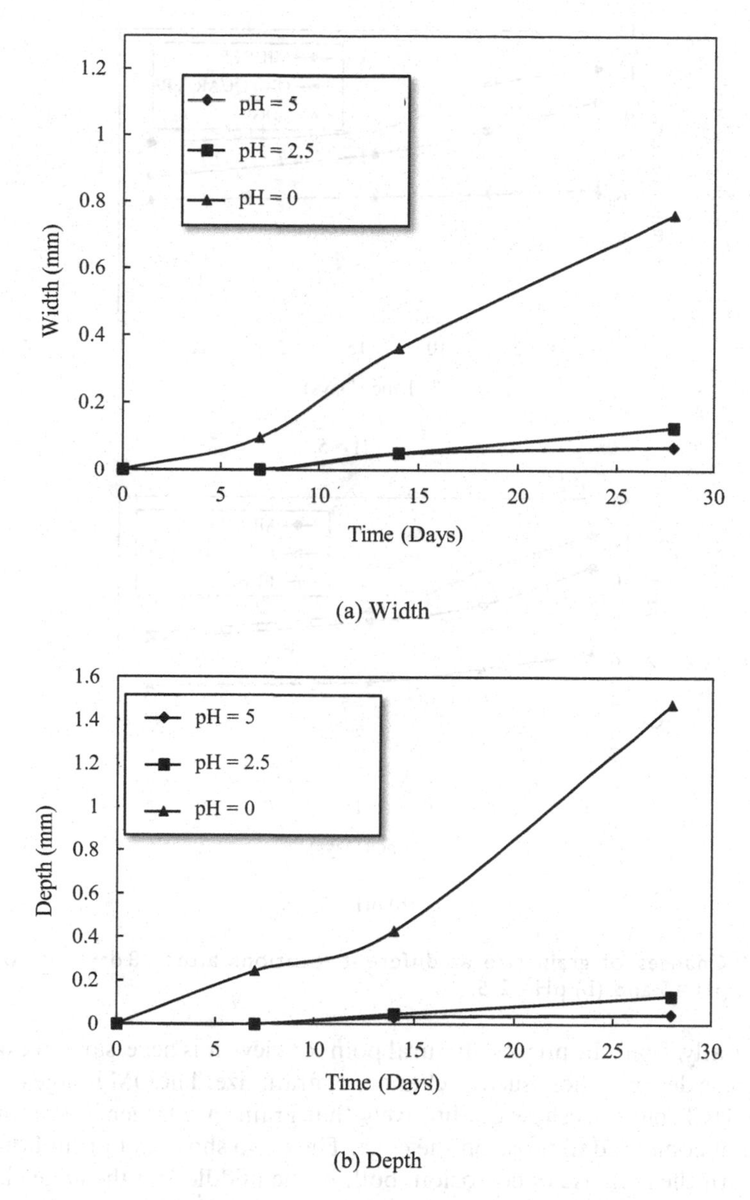

Figure 5.11 **Increase of width and depth of delamination with immersion time in different solutions.**

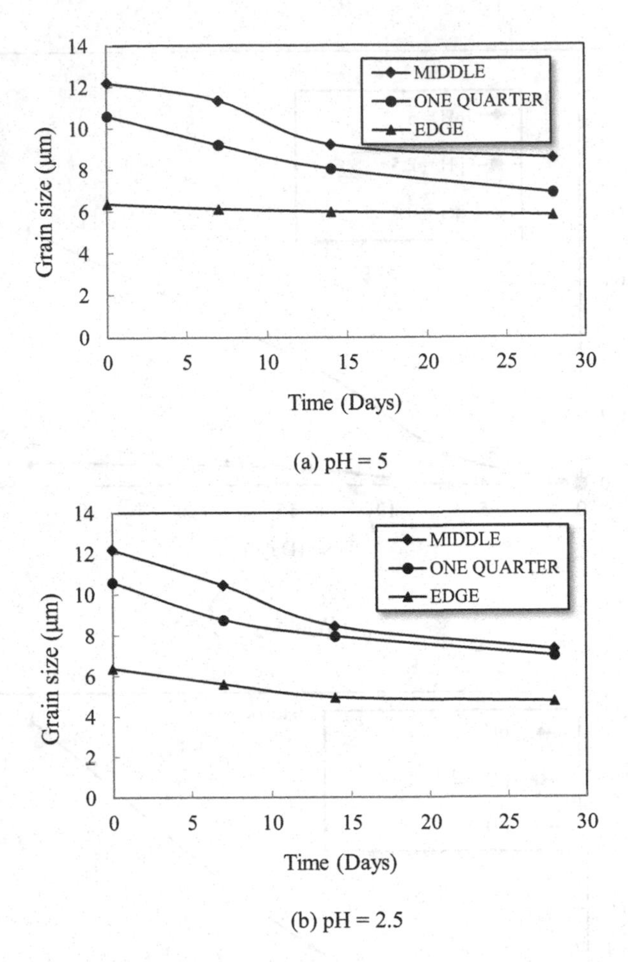

Figure 5.12 Changes of grain size at different locations after 28 days of corrosion: (a) pH = 5 and (b) pH = 2.5.

Fundamentally, from the microstructural point of view, it is necessary to examine what changes cause delamination, such as changes of grain size. The OM images of specimens in various HCl solutions show qualitatively that grains are larger in the middle of the cross section compared to those on the edge. They also show that grain boundaries are widened with the increase of corrosion, both in the middle and the edge close to steel/solution interface, which suggests the reduction in grain size during the corrosion and possibility of cracking. Quantitatively, from Figure 5.12, it can be seen that the average grain size before corrosion is 12.18 μm in the middle of cross section of the samples, 10.60 μm at one quarter of sample cross section and 6.36 μm at the edge. After 28 days of immersion, the grain size in the middle of steel cross section reduces by 29.9% in the solution with pH = 5.0 and 40.1% in the solution with pH = 2.5, respectively. In comparison, the grain sizes at the edge reduce by 9.0% and 25.5%, respectively. Grain size reduces faster in the middle where corrosion is more advanced than at the edge, which leads to more mass loss in the middle and subsequent delamination.

The grain size reduction is due to intergranular corrosion (Sinyavskij et al. 2004). This type of corrosion occurs as a result of more susceptible grain boundaries to corrosion than their centres, due to the depletion of the alloying element (e.g., aluminium and chromium) at the grain boundaries. Intergranular corrosion weakens the bonding force between grains in steel and makes grain boundaries vulnerable to cracking (Parkins 1994). This is the root cause for the initiation of delamination.

The results in the figure also confirm that small grain size at the edge of the steel sample has a higher corrosion resistance than the large grain size in the middle (Marcus 2011, Ralston and Birbilis 2010). This is a result of a passive oxide film being formed before and during corrosion when oxygen attacks the exposed steel atoms; these redox reactions can be described as follows (Marcus 2011):

$$3Fe + 4H_2O \leftrightarrow Fe_3O_4 + 8H^+ + 8e^- \quad (5.1)$$

$$2Fe_3O_4 + H_2O \leftrightarrow 3Fe_2O_3 + 2H^+ + 2e^- \quad (5.2)$$

$$3Fe^{2+} + 4H_2O \leftrightarrow Fe_3O_4 + 8H^+ + 2e^- \quad (5.3)$$

$$2Fe^{2+} + 3H_2O \leftrightarrow Fe_2O_3 + 6H^+ + 2e^- \quad (5.4)$$

The oxide film protects steel from further corrosion. Grain refinement increases the number of grain boundaries, and these boundaries improve the stabilities of passive films since they have a higher energy than the bulk grain (Ralston and Birbilis 2010). In addition, grain refinement improves corrosion resistance by decreasing the compositional difference between bulk grain and grain boundaries, which helps to neutralise the galvanic reactions.

Table 5.3 presents the average iron phase composition at three measured locations across the thickness of steel samples after 28 days of immersion in HCl solutions. As shown in Table 5.3, ferrite (α-Fe) content is around 85% in the middle, 80% at one quarter thickness of the section and 75% at the edge during the entire immersion period (28 days), whilst cementite (Fe_3C) content is around 2% in the middle, 3% at one quarter across the thickness and 5% at the edge. The results indicate that ferrite proportion is higher, and cementite proportion is lower in the middle of the cross section of the sample, compared to that at the edge. There are no significant changes of phase proportion during the corrosion process.

As is known (Marcus 2011), ferrite is corrosion-prone, whilst cementite is corrosion-resistant. It can be seen from the table that in the middle of steel cross section, the ferrite is the highest and cementite content is the lowest. This is the best combination for corrosion. Thus, preferred corrosion occurs at the middle of the cross section. Cementite contributes to corrosion resistance by enhancing the stability of passive

Table 5.3 Quantification of Phase Content (%) of Samples after 28 Days of Immersion

Phase	*Middle*	*One Quarter*	*Edge*
Ferrite (α-Fe)	85	80	75
Cementite (Fe_3C)	2	3	5

films formed during corrosion. This is because the carbon in the cementite improves adherence of the passive oxide film. Although cementite may also cause galvanic reactions that potentially accelerate the corrosion progress, its effect on enhancing adherence of passive oxide film dominates during the corrosion process.

Moreover, higher percentage of ferrite reduces corrosion resistance since the ferrite is more vulnerable to corrosion-induced hydrogen damage than other phases. The existence of hydrogen increases the inner pressure and reduces bond strength between the steel atoms, which reduces the corrosion resistance of the steel. During corrosion, the hydrogen is released and accumulates on the steel surface and gets absorbed into the steel body by diffusion due to a concentration ingredient. The ferrite phase has a higher hydrogen diffusion coefficient than other phases as its crystalline structure favours the residence of hydrogen (Marcus 2011). Therefore, the preferred corrosion progresses faster in the middle and leads to the delamination of steel.

The distribution and composition of impurities is another factor that contributes to the delamination of steel. Impurities can be identified by their formation and precipitation mechanisms. Typically, impurities are formed when oxygen dissolved in the molten steel reacts with an alloying element (aluminium (Al), chromium (Cr) and nickel (Ni)) during the process of steel making. They are then precipitated during steel solidification when their concentration in molten steel increases, which causes impurities to accumulate and reside in the middle of steel, where solidification of the molten steel is slowest and takes the longest time (Thomas 2001).

Figure 5.13 shows the BSE images of specimens after immersion of 28 days in HCl solutions. Compared with Figure 5.8c, it is clear that the impurities (shown as black dots) are corroded away after corrosion. Instead, corrosion pits are formed at the steel/solution interface for corroded specimens. These corrosion pits are likely to be formed due to galvanic corrosion when impurities are exposed to acid (Szklarska-Śmialowska et al. 1970).

The compositions of the observed impurities, as shown in Figure 5.8c, are summarised in Table 5.4. EDS analysis indicates that impurities 1–3 and 5–7 primarily contain oxygen (O) and aluminium (Al), which are likely to be aluminium oxides (Al_2O_3), whilst impurity 4 primarily contains oxygen (O), aluminium (Al), sulphur (S) and manganese (Mn), which are $Al_2O_3 - MnS$ compounds. Impurities accelerate corrosion by inducing galvanic reactions and stress concentration (Marcus 2011). As shown in Figure 5.8c, there are a larger number of impurities in the middle of steel cross section that contribute to preferred corrosion and subsequent delamination of the corroded steel.

5.5 HYDROGEN EMBRITTLEMENT

As discussed in Section 2.4.2, hydrogen embrittlement is a phenomenon whereby hydrogen is absorbed in steel (e.g., by diffuses) to a certain concentration which then exerts local stresses and leads to brittle fracture of the steel. The source of hydrogen is electrochemical reactions of corrosion, i.e., Equation (2.12), from which the hydrogen is released and then trapped in the steel. It should be noted that hydrogen embrittlement is not the only way in which materials are damaged by hydrogen. There are three categories of hydrogen damage: (i) high-temperature hydrogen attack or simply

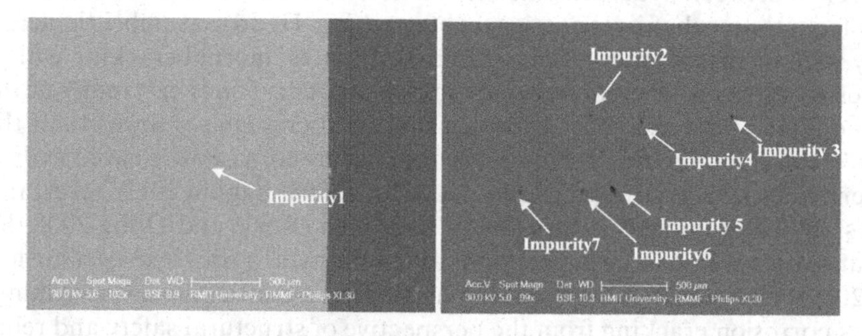

(a) Un-corroded steel

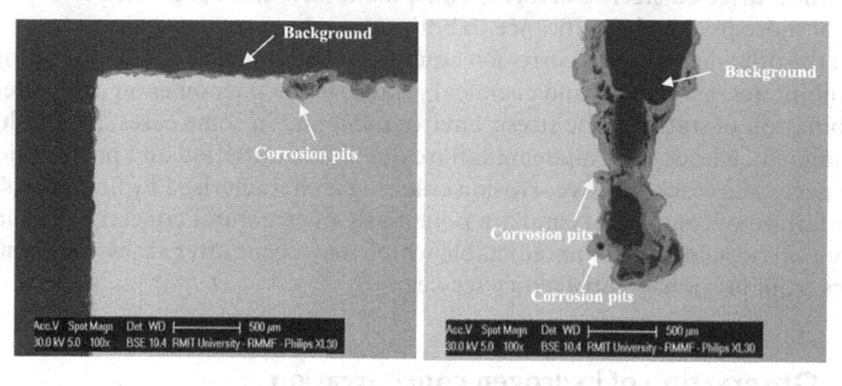

(b) Corroded steel

Figure 5.13 Impurities at the edge (a) and the middle (b) across steel thickness after corrosion.

Table 5.4 Chemical Composition of Impurities

Impurity	*O (%)*	*Al (%)*	*S (%)*	*Mn (%)*	*Fe (%)*	*Others (%)*
1	26.10	35.12	0.00	0.84	37.93	0.01
2	36.02	37.01	0.04	0.00	26.93	0.00
3	26.39	32.96	0.67	1.73	33.63	4.62
4	32.88	38.76	3.07	6.86	18.66	0.00
5	13.74	18.12	0.00	0.36	67.78	0.00
6	11.09	12.89	0.00	0.41	75.04	0.57
7	20.31	28.71	0.17	0.72	49.57	0.52

hydrogen damage, (ii) hydrogen blistering or more precisely hydrogen induced cracking and (iii) hydrogen embrittlement. This section focuses on the third one, i.e., hydrogen embrittlement. The first and second types of hydrogen damages can be referred to in other books, such as Marcus (2011).

It needs also to be noted that the hydrogen embrittlement should not be mixed with stress corrosion cracking simply because they both cause or initiate cracking.

The difference between hydrogen embrittlement and stress corrosion cracking can be identified from the mechanisms or triggers of cracking. Hydrogen embrittlement starts internally with increased internal pressure that initiates internal cracking which then propagates to surfaces, whereas stress corrosion cracking starts externally at the surface. The degree of corrosion for hydrogen=induced cracking is lower than that for stress corrosion cracking. Furthermore, the damage by corrosion-released hydrogen is more severe when steel is stressed (Revie and Uhlig 2008, Eggum 2013). Stress initiates the cracks, which promote the diffusion of hydrogen (Revie and Uhlig 2008, Eggum 2013). Cracks also create more dislocations and voids which trap hydrogen inside steel (Eggum 2013). This can be the reason that hydrogen embrittlement is more dangerous than stress corrosion cracking from the perspective of structural safety and reliability of corrosion-affected steel structures. Thus, the information presented in this section can be of more practical significance to both researchers and practitioners.

On the other hand, stress corrosion cracking is the failure of a metal resulting from the conjoint action of stress and chemical attack. It is a phenomenon associated with a combination of static tensile stress, environment and in some cases, a metallurgical condition which leads to component failure due to the initiation and propagation of a high aspect ratio crack. Stress corrosion cracking is characterised by fine cracks which lead to failure of components and are potentially of structural concerns. The failures are more often sudden and unpredictable which may occur after as less as few months or years from previously satisfactory service.

5.5.1 Observation of hydrogen concentration

Hydrogen concentration can be measured from corrosion tests on steel specimens using a barnacle cell system, which is made and calibrated following ASTM F1113-87 (ASTM, 2017b). In theory, Barnacle cell takes hydrogen-containing steel as an anode and uses a nickel/nickel oxide electrode as a cathode (Li 2018). For the measurements of corrosion, steel specimens are exposed to sodium hydroxide (NaOH) solutions restored in a Teflon cell. Hydrogen atoms in the steel react with hydroxide solutions. The current of reaction is then recorded after 30 minutes to calculate hydrogen concentration, using the following equation (Li et al. 2018a):

$$I_P = F[H]\sqrt{\frac{D_f}{\pi t}} \tag{5.5}$$

where I_P is the current density, F is the Faraday constant (96485.3 C/mol), $[H]$ is the hydrogen concentration, t is the recording time (30 minutes) and D_f is the diffusion coefficient for mild steel (2.5×10^{-8} cm/s^2) (ASTM, 2017a).

A typical measurement of hydrogen concentration in specimens immersed in solutions with various acidities of pH = 0. 2.5 and 5.0 is shown in Figure 5.14. It can be seen from the figure that hydrogen concentration increases with the progress of corrosion. Hydrogen concentration grows to 0.57 ppm in the solution with pH = 5 and 1.58 ppm in the solution with pH = 2.5 after 14 days of immersion and reaches 0.78 and 1.97 ppm in these two solutions, respectively, after 28 days. Moreover, in the solution with pH = 0, hydrogen concentration increases rapidly to 4.47 ppm after 14 days of

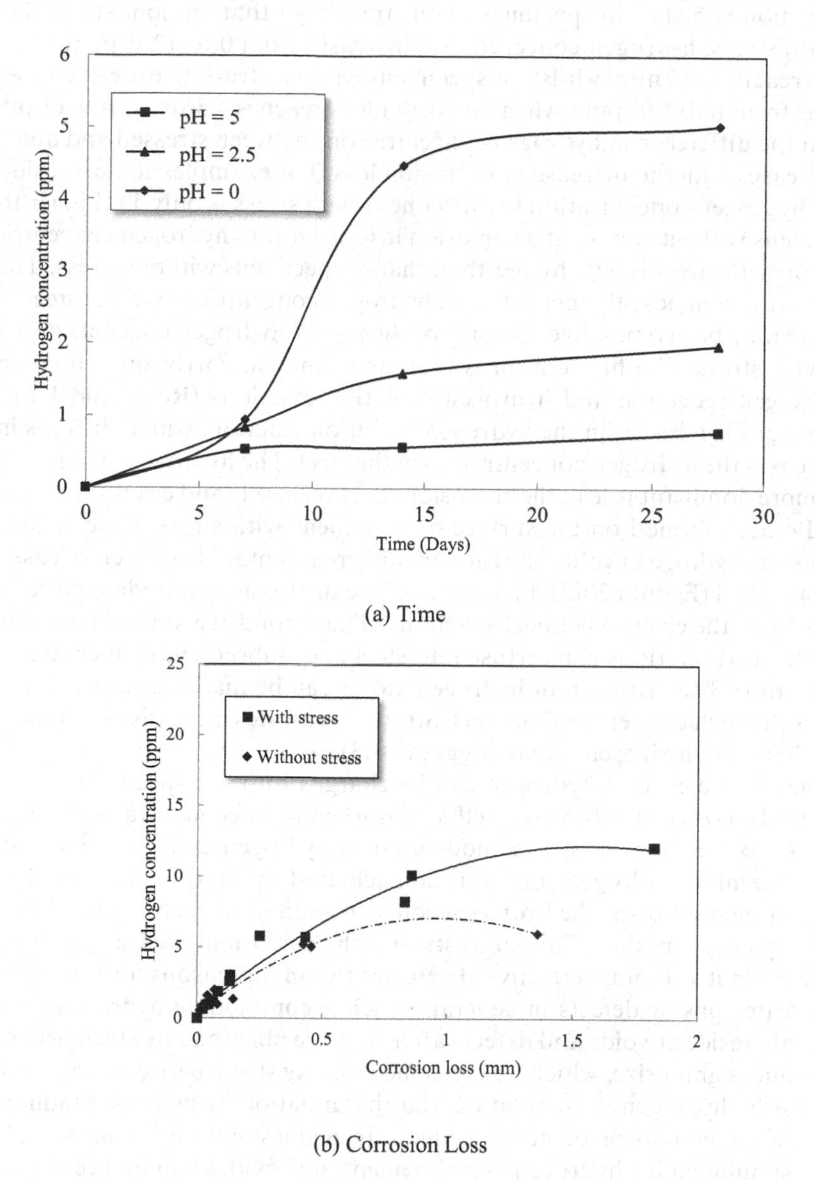

Figure 5.14 **Hydrogen concentration with (a) immersion time and (b) corrosion loss.**

immersion and then climbs slowly to 5.03 ppm after 28 days. Generally, the growth rate of hydrogen concentration is reduced along with corrosion progress. This is because hydrogen atoms accumulate and fill up the voids and defects within steel. Thus, the number of residing places for atomic hydrogen decreases during corrosion. The results of Figure 5.14 are consistent with those published in the literature, e.g., Eggum (2013).

Figure 5.14b shows the hydrogen concentration against the corrosion loss for specimens with and without stress in the HCl solution. It can be seen that the hydrogen concentration is higher in specimens with stress than that without stress. For specimens with stress, hydrogen concentration increases from 0 to 12.10 ppm when corrosion loss reaches 1.82 mm, whilst for specimens without stress, hydrogen concentration increases from 0 to 5.97 ppm when corrosion loss reaches 1.36 mm. It is of interest to note that the difference in hydrogen concentrations between stressed and non-stressed steel increases with the increase of corrosion loss. For example, at a corrosion loss of 0.5 mm, hydrogen concentration for specimens with stress is only 1% higher than that of specimens without stress. At a corrosion loss of 1.0 mm, hydrogen concentration for specimens with stress is 50% higher than that of specimens without stress. This shows that the stress significantly increases the hydrogen concentration in the steel.

There may be two possible reasons for the higher hydrogen concentration in specimens with stress. The first reason is that, as is known, corrosion is a combination of the oxygen reduction and hydrogen evolution reactions (Revie and Uhlig 2008). Hydrogen gas is released in the hydrogen evolution reaction, which diffuses into steel and increases the hydrogen concentration in the steel. The hydrogen evolution reaction plays a more dominant role in the corrosion reaction at pits and cracks. There are more pits and cracks formed on the surface of specimens with stress, which subsequently enhances the hydrogen evolution reaction and creates more hydrogen release and diffusion into steel (Eggum 2013). Pits and cracks can also accommodate more hydrogen released from the electrochemical reactions. The second reason is that hydrogen atoms in the acid solutions can diffuse into steel and, subsequently, increase hydrogen concentration. The diffusion of hydrogen atoms can be made easier by stress since it increases the surface energy of the steel. Stress also creates more dislocations and voids in steel that traps hydrogen atoms (Eggum 2013).

It may be noted that hydrogen can be charged into the specimens, e.g., Eggum (2013) and Hejazi et al. (2016), as well as absorbed by specimens as presented in this section. A comparison of two methods used in hydrogen concentration tests shows that the maximum hydrogen concentration achieved by charging is around 1.84 ppm, whilst in immersion tests, the hydrogen concentration by absorption reaches 5.94 ppm after 28 days of corrosion. This suggests that the corrosion-released hydrogen is not only natural but also more effective. There are two main reasons for that. Firstly, corrosion creates pits or defects in general, which accommodate hydrogen atoms since they mainly reside at voids and defects after they are absorbed by steel. Secondly, corrosion reduces grain size, which weakens the bonding stress between steel grains. This facilitates the hydrogen accumulation and the initiation of hydrogen-induced cracking. Therefore, immersion of steel specimens in acidic solutions is a more realistic and effective simulation for hydrogen embrittlement. It provides a more accurate relationship between hydrogen concentration and corrosion loss as shown in Figure 5.14.

5.5.2 Effect of hydrogen concentration

Once hydrogen is absorbed by steel, its effect on the steel is the same, regardless of the source from where it is absorbed. Gaseous hydrogen and hydrogen released from a cathodic reaction differ from each other in two key respects (Li et al. 2019): (i) cathodic hydrogen is adsorbed on the surface as atomic hydrogen (reduced), whereas gaseous hydrogen is adsorbed in the molecular form, and it then dissociates to form atomic hydrogen, and (ii) the internal pressure produced by the gaseous hydrogen is much lower than that produced by cathodic hydrogen, due to the logarithmic term in the Nernst equation, which converts the E value into an exponent on the hydrogen pressure (Eggum 2013).

Figure 5.15 shows how the tensile properties of steel are affected by hydrogen embrittlement. In general, all important tensile properties of steel, such as yield strength, ultimate strength and failure strain, decrease with the increase of hydrogen concentration. It can be seen from the figure that, in the solution with pH = 5, there is 2.34% reduction in yield strength and 4.17% reduction in ultimate strength when the hydrogen content increases to 0.78 ppm. In the solution with pH = 2.5, the reduction of yield strength and ultimate strength is 2.62% and 4.67%, respectively, when the hydrogen content rises to 1.97 ppm. Furthermore, in the solution with pH = 0, the reductions of yield strength and ultimate strength are 2.85% and 6.17%, respectively, when hydrogen concentration reaches 5.03 ppm. It can be seen that the reduction in yield strength is small and in ultimate strength is moderate. This can be that the embrittlement is more reflected in ductility as indicated by failure strain.

From Figure 5.15c, it can be seen that the reduction in failure strain is 12.34% for steel specimens immersed in solution with pH = 5, 26.16% in solution with pH = 2.5 and 42.66% in solution with pH = 0, respectively. These are significantly larger than that for yield strength and ultimate strength. The results show that the ductility of steel reduces dramatically due to hydrogen concentration, in particular, in more acidic solutions or high degree of corrosion. This is because hydrogen accumulation forms molecular hydrogen in steel, which subsequently leads to inner pressure increment and micro-crack initiations (Marcus 2011, Eggum 2013). The significant reduction in ductility or failure strain of corroded steel is perhaps the reason it is called embrittlement.

The results in Figure 5.15 are compared with the results produced from the hydrogen charging test conducted by Hejazi et al. (2016) with gaseous hydrogen. In their tests, there is 0.50% reduction in yield strength and 3.70% reduction in ultimate strength when steel samples were charged with hydrogen of 1 ppm. In immersion tests with pH = 5 as presented in Figure 5.15, the corresponding reduction is about 2.4% in yield strength and 4.5% in ultimate strength, respectively. The difference in these two methods is quite clear. As it is known, in hydrogen charging tests, there are few or no significant changes in elemental composition and microstructural features that would be otherwise caused by active corrosion. Therefore, hydrogen embrittlement can be underestimated based on the charging test. The results in Figure 5.15 represent a more realistic and accurate estimation on the effect of corrosion induced hydrogen embrittlement.

5.5.3 Mechanism for hydrogen embrittlement

If the strength of steel is larger than the stress in the steel, meaning no failure by strength, hydrogen embrittlement can lead to a brittle fracture of steel as a result of hydrogen adsorption. The ultimate failure is instantaneous fracture. The failure by hydrogen embrittlement is mostly intergranular. The fractured surface of the steel has a crystalline appearance. No definitive mechanism of hydrogen embrittlement has been suggested thus far as shown in the published literature (Li 2018).

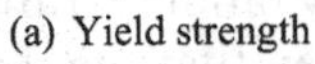

(a) Yield strength

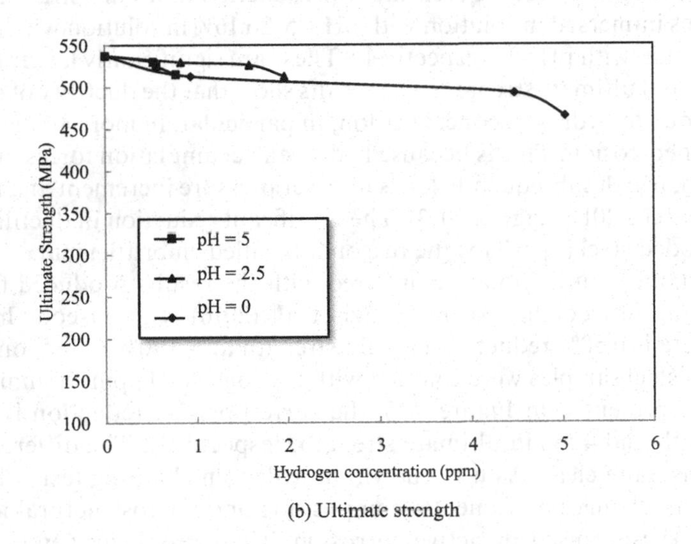

(b) Ultimate strength

Figure 5.15 Reduction of mechanical properties due to hydrogen embrittlement: (a) yield strength, (b) ultimate strength and (c) failure strain.

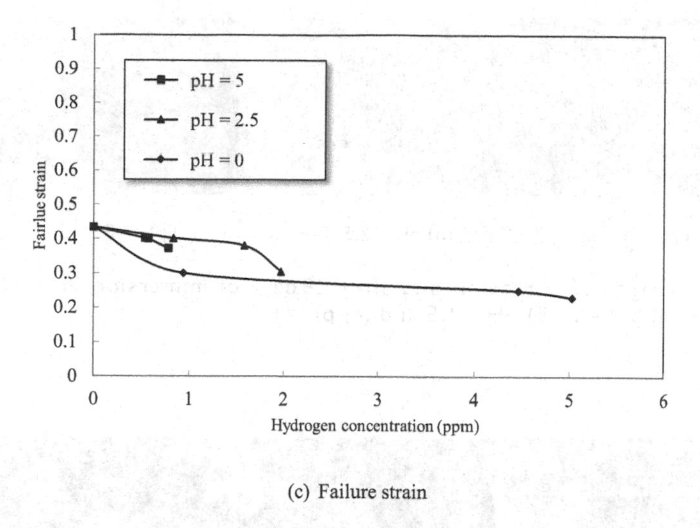

(c) Failure strain

Figure 5.15 (Continued) Reduction of mechanical properties due to hydrogen embrittlement: (a) yield strength, (b) ultimate strength and (c) failure strain.

A generally held view is that impurities segregated at the grain boundary act as catalysts, which increase the adsorption of cathodic hydrogen at these sites. It is widely believed that, in body-central cubic (bcc) iron (α-Fe), hydrogen embrittlement is caused by the interaction of hydrogen with defects in the structure of iron. Such defects include vacancies, dislocations, grain boundaries, interface, voids, etc. Hydrogen is trapped in these defects and facilitates the growth of cracks. A large number of such defects interact with hydrogen and, together with the trapped hydrogen, result in a significant loss of ductility. It has been considered sufficient to identify hydrogen as a cause of cracking. With this in mind, tests are undertaken to provide evidence that there are impurities and defects in the steel specimens which trigger the embrittlement of steel.

Figure 5.16 shows that the largest corrosion pits are observed close to the boundaries of specimens after 28 days of immersion in HCl solutions with various pH values. The width and depth of these corrosion pits can be measured through ImageJ as described in Section 3.5. Specifically, the width of the corrosion pits is 0.07, 0.54 and 0.47 mm in solutions with pH = 5, 2.5 and 0, respectively. The corresponding depth of these corrosion pits is 0.11, 1.17 and 1.73 mm, respectively. It is very clear from the figure that with the increase of acidity, which causes more corrosion, more corrosion pits occur. Collectively, these pits, or in general defects, trigger the brittle failure.

More interestingly, Figure 5.17 shows that microcracks are detected through SEM analysis in specimens immersed in HCl solutions with various pH values. It can be seen that the cracks are formed next to corrosion pits, which are most likely due to stress concentration. The blisters discovered next to cracks are signs of inner pressure increase due to the accumulation of molecular hydrogen within the steel. Furthermore, after 28 days of corrosion, it is found that the hydrogen concentration is 0.78 ppm in the solution with pH = 5, 1.97 ppm in the solution with pH = 2.5 and

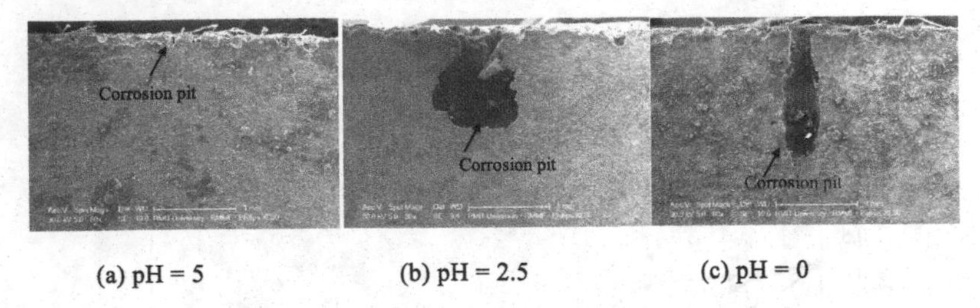

Figure 5.16 SEM images of corrosion pits after 28 days of immersion in acidic solutions with (a) pH = 5, (b) pH = 2.5 and (c) pH = 0.

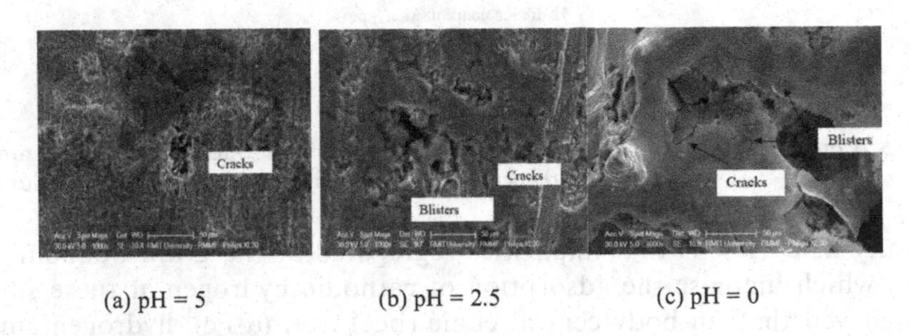

Figure 5.17 Micro-cracks in specimens in solutions with (a) pH = 5, (b) pH = 2.5 and (c) pH = 0.

5.03 ppm in the solution with pH = 0, respectively. Correspondingly, there are more and larger cracks discovered in specimens in the solution with pH = 0 than those in solutions with pH = 2.5 and 5, respectively. These findings suggest that an increase of hydrogen content facilitates the initiation and propagation of cracks. This is mainly because the hydrogen accumulated within steel leads to the increase of residual stress (Li et al. 2019).

Results in Figure 5.17 support the generally held view that micro-cracks are the root causes for hydrogen embrittlement. The mechanism of crack initiation in hydrogen-enriched steel can be one or more of these three models (Li et al. 2019): (i) internal pressure; (ii) hydrogen-induced de-cohesion and (iii) localised slip. In the model of internal pressure, molecular hydrogen accumulates at internal defects (non-metallic inclusions and voids) and builds high internal pressure, which initiates microcracking. In the model of hydrogen-induced de-cohesion, hydrogen concentrates in regions of high tri-axial stress within the steel, which weakens the cohesive force within the steel and makes cracking easier. In the model of localised slip, hydrogen atoms concentrate at defects and weaken the interatomic bonds between irons, which makes steel vulnerable to crack. The tests and results presented in this chapter show that intergranular corrosion occurs along grain boundaries, which reduces the atomic force along grain boundaries and makes grain boundaries vulnerable to cracking. Therefore, to prevent steel degradation due to corrosion, it is essential to prevent intergranular corrosion.

5.6 SUMMARY

Observations of the stress effect on corrosion and degradation of mechanical properties of corroded steel are presented in this chapter. Test results and their analysis show that the stress increases the corrosion significantly and consistently for a range of corrosive environments. It also reduces the tensile properties of corroded steel with most reduction in ductility as measured by failure strain. Causes for preferred corrosion are discussed in this chapter with microstructural analysis to confirm the difference in microstructural features at the edge and in the middle across the thickness of steel cross section, which causes preferred corrosion. Factors that affect preferred corrosion are discussed, and suggestions to prevent preferred corrosion are proposed. Also, in this chapter, simulated corrosion tests are presented to observe preferred corrosion-induced delamination of steel, which is one of the most severe damages of corrosion to steel. With test data, the width and depth of preferred corrosion-induced delamination are quantified, and the mechanisms of delamination are discussed. It is found that larger grain size, higher ferrite content and more impurities in the middle layer of the cross section of steel than those at the edge are the causes for corrosion-induced delamination. Another severe form of corrosion damage is hydrogen embrittlement, which is also covered in this chapter, including measurement of hydrogen concentration, effect of hydrogen embrittlement on tensile properties of corroded steel and the mechanism for hydrogen embrittlement. Test results show that high hydrogen content directly reduces the tensile properties of steel with most reduction in ductility, ultimate strength next and yield strength the least.

Chapter 6

Practical application and future outlook

6.1 INTRODUCTION

Theory is useful only when it is applied in practice. In Chapters 3–5, considerable information and knowledge on corrosion and its effects on degradation of mechanical properties of ferrous metals are presented, based on observations and analysis from experiments and field inspections. How to apply this knowledge to practical assessment of steel and steel structures (or ferrous metals in general) is a question facing all stakeholders of corrosion research and practice. This question is to be addressed in the chapter. It needs to be noted that there is no "one size fits all" approach or method. The methodology and examples presented in the chapter are illustrative and applied to the circumstances as presented. When circumstances are different, modification needs to be made accordingly to the methodology and models presented in the chapter. For this reason, only a general procedure is proposed for readers to take up.

Research experience over 30 years shows that corrosion is a "black hole". There has been so much research conducted on this topic, and yet, there is still so much unknown and hence so much to explore. It would be inaccurate to think that one book has or can cover all the knowledge on corrosion from both the science perspective, such as electrochemical reactions of corrosion, and the engineering perspective, such as corrosion effect on the mechanical properties of steel, in particular, the prediction of degradation of mechanical properties of corroded steel. Whilst it is fully acknowledged that corrosion science lays the foundation of corrosion engineering, it is the corrosion effect on degradation of mechanical properties that matters to the real world of steel and steel structures. How to accurately predict the corrosion effect on the degradation of mechanical properties of corroded steel is and will continue to remain a serious challenge to all stakeholders in corrosion research and practice. This chapter tries to serve the purpose of bringing about more new findings from corrosion research and hence to promote more applications to practical structures.

It is also acknowledged that corrosion is a well-trodden topic with a long history. However, steel continues to corrode, and steel structures continue to collapse, as evidenced in Chapter 1. An examination of such collapses reveals that, surprisingly or not, current collapses are caused by the same corrosion that was known 100 years ago. The reoccurrence of collapses convincingly demonstrates that existing research on corrosion has missed a critical point – how steel changes with corrosion intrinsically from its fundamental atomic structure, i.e., atomic lattice. It is the structure of atomic lattice of steel that determines its property, i.e., strength. This major gap in knowledge hinders the progress of corrosion science, hampers the confident use of steel in structures, and leads to unexpected collapses of corrosion-affected structures. This chapter will shed some light on future research in this direction with a view to achieving accurate prediction of corrosion and its effect on degradation of mechanical properties of steel.

6.2 CALIBRATION OF SIMULATED TESTS

The test results presented in previous chapters are mostly produced from the simulated corrosion tests. These results cannot be applied directly to practical structures for design and assessment of corrosion-affected steel structures although they can provide useful knowledge on how corrosion behaves and what effect it incurs on the degradation of mechanical properties of steel, at least in a qualitative and relative manner. To apply these results and knowledge derived from them, a calibration process is essential. The simulated corrosion tests are conducted in an artificial environment in which the corrosion process and the effect of corrosion on mechanical properties are, understandably, different from those in natural and real environments. How to achieve this calibration is the topic of this section.

Since corrosion-induced damages to steel and steel structures take a long time to manifest and cause significant consequences, acceleration is one of most used methods for corrosion tests. The applicability of data produced from the accelerated tests is well recognised, but the solution is yet to be developed. One such solution is discussed in the section.

6.2.1 Basics of similarity theory

Similarity theory is a tool used in engineering to measure the resemblance between two sets of data or two models, to validate the applicability of one set of data to another or one model to another. In the context of this chapter, and for the purpose of calibration, two sets of data are selected such that one is from tests in an accelerated environment, referred to as a model here, and the other is from tests under the natural environment, known as prototype. As for almost all research, models are typically smaller than the prototype or actual components of structures. It is therefore important that the models can be designed to share the essential similarities with a prototype, i.e., a real application. To achieve similitude between test model and prototype, three fundamental similarity requirements must be satisfied (Hubert 2009). The first is geometric similarity. Model (m) and prototype (p) are similar if both of them have the same shape and their dimensions are related by a constant scaling factor. This means that two objects are congruent to the result of scaling, rotating and repositioning (Yan and Li 2015). Geometric similarity can be mathematically presented as follows:

$$D_m \,/\, D_p = s_g \tag{6.1}$$

where D denotes the dimension, the subscripts m and p represent model or prototype, respectively, and s_g is a constant ratio, known as the scaling factor for geometric similarity. A constant scaling factor means the ratio of any dimension of the model to that of the prototype is the same. Therefore, objects with geometrical similarity, such as model and prototype, are similar in shape and proportional in physical dimensions although they differ in actual sizes. This kind of objects includes spheres, cubes and regular tetrahedra.

The second similarity requirement is kinematic similarity, which emphasises on similarity of motion of the objects. Since motion is related to distance and time, it refers to the similarity of lengths (i.e., geometric similarity) and similarity of time intervals. If the length scaling factor is s_g given by Equation (6.1) and the ratio of the corresponding time intervals is s_t, i.e., scaling factor for time intervals, then the scaling factor for motion, i.e., velocity, can be defined as the ratio s_g / s_t, and the scaling factor for acceleration can be defined as the ratio s_g / s_t^2. It can be seen that the geometric similarity is a necessary condition for the kinematic similarity but not the sufficient one. The third similarity requirement is the dynamic similarity, which is related to forces. It requires that identical types of forces on two objects, such as model and prototype, are related in magnitude and direction by a fixed ratio, i.e., dynamic scaling factor. Forces are not only those loads, such as pressure and gravity, but also mechanical properties, such as stress and strain. For example, let ε be strain, σ be stress and E modulus of elasticity of an object, such as model and prototype; they have to satisfy the following similarity requirements:

$$\varepsilon_m = \varepsilon_p, \sigma_m / \sigma_p = s_d, \; E_m / E_p = s_d \tag{6.2}$$

where s_d is a constant ratio, known as the scaling factor for dynamic similarity. It may be noted in Equation (6.2) that strains in model and prototype are the same. This is because strain is relative change in length, the dimension or "scale" of which is cancelled.

Similarity theory provides a quantitative relationship for parameters between the model and prototype. It needs to be noted that the definition of similarity theory depends heavily on experiences. Obtaining a satisfactory outcome of similarity depends on the experience of the specific problem. Also, each similarity requirement is closely tied to a particular application (Cao and Lin 2008, Yan and Li 2015). Not all similarity requirements can be satisfied, and some may be satisfied only under specific circumstances. Therefore, they cannot be used interchangeably or universally.

To overcome these difficulties, an objective function is introduced to determine the interrelationship between two sets of data from, such as model and prototype. An ideal objective function should be derived from relevant physical processes and rigorous mathematics. The accuracy of the objective function should be verified by in-situ experiments. Also, the objective function needs to be universal and can solve multi-objective problems under the same scale. It is better to be unitless for normalisation (Yan and Li 2015).

6.2.2 Acceleration factor

The essence of similarity theory is the scaling factor, and in the context of accelerated corrosion, it is to scale the data produced under the accelerated conditions or environments to that under natural conditions. Based on the concept of similarity theory described in Section 6.2.1, the acceleration scaling factor can be introduced to relate the two sets of data. As it is appreciated, data acquired under accelerated conditions for corrosion in simulated solutions has to be translated to that under the conditions typical of service to be practically useful and usable. To address and verify the applicability of the data (and subsequent models) acquired under the accelerated conditions to conditions typical of service, the nature of the acceleration needs to be examined. In accelerated corrosion tests, it is the time that is needed to produce a certain physical effect, such as corrosion loss, that is accelerated. It follows that the nature of acceleration is to shorten the time that actually is required to produce the physical effect. Thus, as first proposed in Li (2000), the key to translate the test data from accelerated corrosion tests to that under the conditions typical of service is to translate the "accelerated" period of time, which is the "test" period of time it takes for a given physical effect, to a "real" period of time, which is the equivalent "natural" period of time it takes for the same given physical effect. This concept is known as time transformation in this chapter. Time transformation can only be achieved through calibration of accelerated tests against (long-term) tests under a designated service condition. The principle of time transformation is to gauge the equivalence of accelerated time to natural time. In theory, this gauge is controlled and determined by similarity requirements as described in Section 6.2.1. In practice however, not all requirements can be strictly met.

The basic concept of time transformation is as follows. Tests on identical specimens are carried out in two sets of environmental conditions. The identical specimens meet the geometric similarity requirement. One is in an accelerated condition for corrosion, which is a simulated condition, such as acidic solution in Sections 3.2 and 4.2. The other is under the natural ambient condition representing a typical service condition. With these two sets of conditions and subsequent two sets of data on the intended physical effect, the essential problem is to determine the equivalent time period that induces the given physical effect under the accelerated conditions to the real time period that induces the same given physical effect under the natural conditions. This meets the dynamic similarity requirement. To achieve this, some typical parameters can be selected to measure and determine the physical effect or parameter induced by corrosion. Ideally, one is the corrosion loss and strength reduction as external or macrophysical effect, and the other is element content and grain size as internal or microphysical effect. The principle is that at the end of given time under both accelerated and natural conditions, the measured physical effect, i.e., corrosion loss, strength, element content and grain size, from both sets of identical specimens under two sets of environmental conditions should be exactly the same. The natural period of time is then the equivalent period of time of the accelerated time. In practice, however, as experienced in the tests presented in the book, measurement of these selected parameters from the two sets of specimens rarely matches at the same time under the two sets of conditions. Thus, the final equivalent time has to be based on the mean value of more sets of data from the two sets of identical specimens.

It is acknowledged that the measurement of a physical effect may not have considered the chemical process of corrosion at an instantaneous point in time; in particular, the electrochemical reactions are dynamic. This can be important if the electrochemical reactions are taken into account in the measurement of the physical effect. As stated from the outset, this book takes a phenomenological approach in observing the corrosion process with focus on resultant physical effect from the mechanistic perspective. This essentially defines the corrosion effect by physical parameters rather than chemical reactions, focusing on end result rather than the process. Also in this book, observation of corrosion-induced changes of mechanical properties is in relative terms rather than in absolute values. Thus, end results rather than the process can be sufficient in representing the effect of corrosion. Furthermore, the intention of this chapter is to propose a method that can achieve the time transformation so that the long-term tests on corrosion and its effect on mechanical properties of steel can be carried out in a relatively short period of time, which makes a very difficult test possible. The accuracy of this method obviously depends on the accuracy of the control of two sets of environments, which can be achieved, such as, in two large environmental chambers. Obviously, once the resources are available, accurate time transformations can be derived.

It is also acknowledged that deficiencies may exist in this method of experimental calibration, e.g., ambient conditions or typical service conditions are usually not as constant as required so that different ambient conditions may result in different time periods of producing the same physical effect. This deficiency, however, is not methodological and can be overcome since the accuracy of the experimental calibration depends on the accuracy of the control of two sets of environments. It would be ideal to have two sets of controllable conditions, such as those in two large chambers. One is the accelerated condition and the other represents a designated real service condition to which the accelerated test data are to be applied. Thus, as long as resources are available, accurate calibration can be achieved.

As first proposed by Li (2000), an acceleration factor can be introduced to quantify the time transformation. The acceleration factor, denoted by s_t, is a scaling factor for time, which is defined in this chapter as the ratio of two change rates of the same physical effect in two sets of environments and expressed as follows:

$$s_t = \frac{\gamma_a}{\gamma_n} \tag{6.3}$$

where γ_a is the change rate of a given physical effect induced by corrosion in the accelerated corrosion environment, and γ_n is the change rate of the same physical effect induced by corrosion in a natural service environment. Obviously, s_t is unitless, which meets the requirement of similarity theory. Equation (6.3) is an empirical formula that can be used to estimate the acceleration factors for the test specimens in various accelerated environments to natural environment as to be shown in the examples of the next section.

It should be noted that an implicit assumption in Equation (6.3) is that the change of physical effect is linear, otherwise the acceleration factor is not constant. This is not true in reality, but for some simplification, linearity can be assumed. Obviously, the linear assumption does not affect the concept and its application. It should also be noted that even if the acceleration factor cannot be derived with a reliable accuracy at the moment, the results from accelerated tests are still of merit in the view that qualitative, comparative and sensitivity studies on the effect of different factors and parameters affecting the steel corrosion on corrosion and their effects on degradation of mechanical properties of corroded steel can be carried out, as have been demonstrated in the previous chapters. Thus, the practical significance of the accelerated test and test results should not be underestimated.

6.2.3 Examples

Equation (6.3) can be used to translate the test data from the accelerated conditions or environments to that of natural conditions typical of service environments. Using the information presented in Chapters 3–5, acceleration factors for selected physical effects in various test conditions can be determined as shown in the examples below.

Example 6.1: Corrosion loss of steel

In this example, let the physical effect be corrosion loss in mm of steel due to corrosion. Assume the first set of environments to be acidic solution with pH = 5.0. From Figure 3.3 of Chapter 3, the relationship of corrosion loss with the accelerated time (immersion time) can be expressed as follows:

$$C = 0.0013t + 0.0016 \tag{6.4}$$

where C is corrosion loss in mm, and t is the time in accelerated days. The coefficient 0.0013 is the change rate of corrosion loss, i.e., corrosion rate.

Let the environment where the test results are to be applied be atmosphere. From Figure 3.6 of Chapter 3, the relationship of corrosion loss with the natural time can be expressed as follows:

$$C = 0.00002t + 0.036 \tag{6.5}$$

where C is the same corrosion loss in mm, but t is the time in natural days in atmosphere. From Equation (6.3), the acceleration factor is

$$s_t = \frac{\gamma_a}{\gamma_n} = \frac{0.0013}{0.00002} = 65 \tag{6.6}$$

Thus, to apply test results from this immersion test in the acidic solution with pH = 5.0 to corrosion in the atmospheric environment, the acceleration factor of 65 can be applied.

Chapter 3 also presented test results produced from other acidic solutions with pH = 0 and 2.5. A summary of acceleration factors for various acidic solutions to atmospheric corrosion is presented in Table 6.1.

Two points should be noted. One is that the acceleration factor should not be constant due to non-linearity of the corrosion progress. Thus, the acceleration factor can

Table 6.1 Acceleration Factor for Atmospheric Corrosion

Accelerated Environment	*Specimen without Stress*	*Specimen with Stress*
HCl solution with pH = 5.5	65	100
HCl solution with pH = 2.5	120	225
HCl solution with pH = 0.0	815	1,545

be regarded constant only for a time interval in which the corrosion rate is linear. The other point is that since a large degree of variation has been experienced in deriving the acceleration factor, it may be an illusive task to determine an accurate acceleration factor based on the test results currently available. More calibration tests under more rigorous (or controllable) environmental conditions are necessary. Therefore, the acceleration factors in Table 6.1 are somehow indicative, derived from the "filtered" data with coefficient of variations less than 0.5.

Example 6.2: Reduction of modulus of rupture

The calculation of acceleration factor for corrosion loss of cast iron is the same as that in Example 6.1 for steel corrosion. Thus, in this example, let the physical effect be the modulus of rupture (strength) of cast iron (pipe) section. Assume the first set of environments to be soil with pH = 2.5 and 80% saturation. From Figure 4.13a of Chapter 4, the relationship of the relative reduction of the modulus of rupture with the accelerated time (burial time) can be expressed as follows:

$$\Delta M_s = 0.013t \tag{6.7}$$

where ΔM_s is the relative reduction of modulus of rupture in %, and t is the time in burial or accelerated days. The coefficient 0.013 is the change rate of the reduction, i.e., reduction rate.

Let the environment where the test results are to be applied be natural soil. From Figure 4.13b of Chapter 4, the relationship of the relative reduction of the modulus of rupture with the real natural time can be expressed as follows:

$$\Delta M_s = 0.0011t \tag{6.8}$$

where ΔM_s is the relative reduction of the same modulus of rupture in %, but t is the time in natural days in natural soil. From Equation (6.3), the acceleration factor is $s_t = 12$ (=0.013/0.0011). Thus, to apply test results produced from this accelerated soil with pH = 2.5 and 80% saturation to corrosion-induced reduction of modulus of rupture in natural soil, the acceleration factor 12 can be applied. For example, for a given percentage reduction of modulus in a unit time in accelerated soil, it will take 12 units of natural time for corrosion to induce the same percentage reduction of modulus of rupture.

For test data from acidic solution with various pH values as presented in Chapter 3–5, corresponding acceleration factors can be determined and populated as those in Table 6.1. This is straightforward and hence omitted here.

Example 6.3: Change of element content

In Examples 6.2 and 6.3, external and macroparameters, i.e., corrosion loss and modulus of rupture, are used for determining the acceleration factors. In this example, it is shown that internal and microparameters can also be used for determining the acceleration factor based on Equation (6.3). Now, let the physical effect be the iron content in steel. Assume the first set of environments to be simulated soil solution with pH = 5. From Figure 3.24a of Chapter 3, the reduction rate, or change rate, of iron content in steel with the accelerated time (immersion time) is 0.00048% per accelerated day.

Let the environment where the test results are to be applied be natural soil. From Figure 4.23a of Chapter 4, the reduction rate, or change rate, of iron content in steel with the natural time in natural soil is 0.00003% per natural day. From Equation (6.3), the acceleration factor is s_t = 16 (= 0.00048/0.00003). Thus, to apply test results produced from this simulated soil solution with pH = 5 to corrosion-induced reduction of iron content in steel buried in natural soil, the acceleration factor of 16 can be applied. It should be pointed out that linearity of iron content reduction during corrosion is assumed for both sets of environments for simplicity of the illustration. In reality of course, it is not linear, and as such, more complicated calculation is required, but the principle or concept of time transformation remains the same.

It may be noted from these three examples that the acceleration factors for different physical parameters can be different. In general, corrosion in the atmosphere is much slower than in soil as reflected in the acceleration factors, i.e., 65 in the atmosphere and 12 and 16 in soils. This can be some sort of vindication of the concept of time transformation presented in Section 6.2.2. Even in the same sets of environments, the acceleration factors can be different for different physical effects or parameters. This can be understandable since not all physical effects or parameters induced by corrosion change proportionally with corrosion and/or with each other. Having said that, it is acknowledged that the method of acceleration factor is empirical and hence approximate when data are not sufficient. The significance of the method lies in its principle and concept rather than its accuracy, which depends on sufficient data as all other empirical methods would do.

6.3 PRACTICAL APPLICATIONS

Application of data presented in the book to practical design and assessment of corrosion and degradation of mechanical properties of steel and steel structures in service is one of the primary aims of this book. To achieve this, a general procedure for practical application for design and assessment is presented, which can be applied to corrosion-affected or corrosion-prone steel and steel structures in various environments. Two examples are presented to illustrate how to apply the test results presented in previous chapters to the design and assessment of steel structures with more focus on assessment. Corrosion of both steel and cast iron and degradation of their mechanical properties are covered. The examples can serve as a guide for practitioners in design and assessment of corrosion of steel and structural deterioration of steel structures caused by corrosion.

6.3.1 General procedure

Compared with structural assessment, structural design is quite mature as evidenced by codes and standards widely available that are issued and governed by national authorities in almost all countries. Thus, full structural design will not be covered in this section except that the degradation of material properties is not included in most of the codes and standards which deserve some effort in this section. A general procedure to include degradation of mechanical properties of steel in design of steel structures can be summarised as follows:

1. Determine the environment where the structure is to be located and operated;
2. Determine an acceptable sacrificial loss of thickness of structural members, i.e., corrosion loss in mm;
3. Select an accelerated environment presented in Chapters 3–5;
4. With the given sacrificial corrosion loss, determine the acceleration factor from the selected environment to the service environment, using the method presented in Section 6.2.2;
5. Determine the relative reduction of mechanical properties from relevant figures presented in Chapters 3–5 by multiplying them by the acceleration factor determined in step 4;
6. For a given design service life, t_L, the maximum reduction of mechanical properties can be determined, which is then used in design calculations. Vice versa, for a given acceptable reduction of mechanical properties, the design service life can be determined.

In comparison, assessment of corrosion and in particular its effect on steel and steel structures is less standardised and more complicated. There are many procedures for corrosion assessment on steel and steel structures with perhaps most focusing on condition assessment rather than structural assessment. The difference between these two assessments mainly lies in perspective. The condition assessment is more from materials perspective, such as rusts, corrosion pits, cracks and superficial damages and something like these. The structural assessment is more from the perspective of structural functions, such as safety and serviceability with clear and mandatory requirements and criteria stipulated in regulatory codes and standards issued by an authority of a country. Even in condition assessment or structural assessment, there are many procedures serving for different purposes. It is believed that the principle of assessment in all these procedures is more or less the same. Bearing this in mind, this chapter outlines an assessment procedure that is, in principle, most suitable for structural assessment for corrosion-affected steel structures. The purpose of this procedure is primarily to apply the corrosion data presented in previous chapters to practical structure so as to maximise the value of these test data. A general procedure for assessing a corrosion-affected steel structure can be outlined as follows:

1. Collect design and construction information of the structure, including drawings, materials used, loadings applied, any defects or damages due to fabrication, such as residual stress due to welding, or any measures for corrosion protection, such as coating, cathodic protection, etcetera;
2. Collect information about any maintenance of the structure, including time, such as year, methods, such as repair or strengthening, and materials used, such as concrete;

3. Collect information about the environment in which the structure is exposed, including, seasonable changes of temperature and humidity, any aggressive agents in the air, such as salts, sulphates and any astray currents;
4. Visual inspection of the structure, recording any defects, discrepancies from design and construction drawings, in particular corrosion loss, corrosion pits and other damages;
5. Analysis of inspection data to decide or estimate or categorise the environment for corrosion, such as atmosphere or marine, types of corrosion, such as uniform corrosion or pitting corrosion, and actual corrosion loss;
6. Gauge the actual environment to an accelerated environment for corrosion as much closely as possible. If this is not possible, it should be considered to take samples from the structure for necessary corrosion tests and corresponding mechanical testing;
7. Correlate the actual corrosion loss in mm to that from accelerated tests, such as using time transformation method presented in Section 6.2 or other appropriate methods;
8. Estimate the degradation of mechanical capacity of the material (i.e., ferrous metals) due to corrosion, including tensile strength, fatigue strength and fracture toughness, such as using data presented in previous chapters or other appropriate data;
9. Estimate the structural capacity based on criteria specified in regulatory design codes and standards, such as ultimate and serviceability limit states, and follow the clauses and formulae provided in the codes and standards.

The key steps in this procedure in relation to assessment of corrosion-affected structures and application of test data presented in this book are Steps 6–9. These four steps are for structural assessment, which are quite different from condition assessment and hence will be explained further using specific examples in the next section.

6.3.2 Steel structures

Example 6.4: Design

This simple example illustrates how to design a steel structure with consideration of degradation of mechanical properties of steel due to corrosion. For illustration purposes, all other design parameters, such as loading, dimensions and so on, are assumed given and not changed during the design service life of the structure. For a more comprehensive design for structures in their whole life of service, considering the deterioration of materials and structures, publications of the authors can be referred to, such as Li (2004), Yang et al. (2018) and Wang et al. (2019b), to name a few.

Now design a steel structural member subjected to sustained service loads and corrosion for tension failure as a design criterion, which is the ultimate limit state. The structure is located in an industrial area with some corrosive pollutants in the air. For this reason, a sacrificial corrosion loss of 0.01 mm/year is accounted for the dimension of the structural member in its design due to expected corrosion.

To make use of test data on steel corrosion and its effect on degradation of tensile strength as presented in previous chapters, an acceleration factor needs to be determined as per Steps (3) and (4) of the general procedure outlined in Section 6.3.1. Since the structure is located in an industrial area, test data produced from the acidic solution with pH = 5 can be the closest to this service environment. Also, since the structure is subjected to sustained service load, test data on corrosion from specimens immersed in the acidic solution with stress are used. From Figure 5.3a of Chapter 5, the corrosion rate for steel in such environment is 0.0021 mm/accelerated day, i.e., immersion day. The sacrificial corrosion loss of the dimension or thickness of the structural member in the service environment is designed to be 0.01 mm/natural year, i.e., 0.000027 mm/natural day. From these two sets of data, the acceleration factor can be determined which is $s_t = 77$ (= 0.0021/0.000027). This acceleration factor can be used to gauge the accelerated time to real natural time as an estimation.

The relationship between the relative reduction of ultimate strength and immersion time in the solution with pH = 5 can be obtained from Figure 3.12a of Chapter 3 from regression analysis, which is expressed as follows:

$$\sigma_u(t) = \sigma_{u0}\left(1 - \frac{0.15t}{100}\right) \tag{6.9}$$

where σ_{u0} and σ_u are the original ultimate strength and ultimate strength at time t in immersion or accelerated days, respectively. The coefficient of determination for Equation (6.9) is 0.99. For a given acceptable reduction of 10%, such as a resistance factor of 0.9, as specified in codes and standards, from Equation (6.9)

$$\frac{\sigma_u(t)}{\sigma_{u0}} = \left(1 - \frac{0.15t}{100}\right) = 0.1 \tag{6.10}$$

it can be obtained that t is 600 accelerated days. Using the acceleration factor of 77, it is equivalent to 126 $\left(= \frac{600 \times 77}{365}\right)$ natural years. Since the ultimate strength is the design criteria, the design service life for this steel structural member under the ultimate strength criterion can be estimated as 126 years.

On the other hand, for a given design service life of 100 years, the maximum reduction of the ultimate strength of the structural member increases to about 25%. This means that if the mechanical strength of the steel is expected to reduce more due to corrosion, the design service life is reduced. In other words, more reduction of ultimate strength can be accepted but at the cost of shortened service life. This calculation is straightforward, the details of which are omitted here. It should be noted that this example is purely for illustration of how to apply test data presented in this book to design of corrosion-affected or corrosion-prone structures. There are many assumptions acquiesced in this example, such as linear corrosion and linear reduction of ultimate strength. As discussed in previous chapters, this is only technical and not methodological, which is the purpose of this example.

Example 6.5: Assessment

It may be appreciated that assessment of corrosion-affected steel structure is more complicated than its design, not to mention that not codes and standards can be followed. Now consider a simply supported steel girder bridge across a sea estuary with a span of 16 m. The bridge was constructed 90 years ago and subjected to marine corrosion and traffic loads. The section of the girder is a rectangular hollow section with a dimension of 600 (height) mm by 200 (width) mm. The thickness of the section wall is 10 mm. The original yield strength of the steel is $\sigma_{yo} = 310$ MPa before corrosion. The beam is subjected to dead load and live load. The dead load is 1.5 kN/m in total. The live load is also uniformly distributed, but its value fluctuates between 0 and 2.5 kN/m with the frequency 25 cycles/day. Due to long-term service in a corrosive environment, it is required to assess the remaining structural capacity of the steel girder to ensure the safe and serviceable operation of the bridge.

Following the procedure outlined in Section 6.3.1, it is assumed that Steps (1)–(3) have been undertaken as described above. From Step (4), a visual inspection needs to be carried out to know the corrosion damage to the steel as measured by corrosion loss. Site inspection shows that the corrosion is uniform and can be assumed to occur around the perimeter of the rectangular hollow section. The average corrosion loss in thickness as measured in situ after cleaning the rusts is 0.3 mm. Since the bridge is still in service, it is not possible to determine its corrosion rate, i.e., actual corrosion loss over time. It is assumed that corrosion began in the beginning of its service.

The concept of time transformation is used to gauge the actual environment to an accelerated environment as per Step (6) of the general procedure. Using data from immersion tests in HCl solution with pH = 5 (the closest to marine environment from all tests data), 0.3 mm in 90 years in a natural environment is equivalent 230 days in accelerated time, i.e., immersion time in the HCl solution, as determined from Equation (6.4). With this equivalence, the acceleration factor is 143 (=90*365/230), as determined from Equation (6.3). It should be noted that, strictly speaking, the acceleration factor should be determined from the change rate of the measured parameters, but in this example, it is not possible to estimate the change rate of corrosion, i.e., corrosion rate, accurately since there is only one data point on corrosion loss.

With the equivalent time in immersion tests, the residual strength or remaining strength of the corroded steel can be determined accordingly from the test results presented in Chapter 3. Two types of strength of steel are considered here in the assessment. One is the yield strength of the steel, and the other is fatigue strength. Following Equation (6.9), the relationship between the relative reduction of yield strength and immersion time in the solution with pH = 5 can be obtained from Figure 3.8a of Chapter 3 from regression analysis, which is expressed as follows:

$$\sigma_y(t) = \sigma_{y0}\left(1 - \frac{0.08t}{100}\right) \tag{6.11}$$

where σ_{y0} and σ_y are original yield strength and yield strength at time t in immersion or accelerated days, respectively. The coefficient of determination for Equation (6.10) is 0.91. From Equation (6.11) and with $t = 230$ accelerated days (90 natural years) and

$\sigma_{y0} = 310$ MPa, the residual yield strength can be determined as $\sigma_y = 253$ MPa. This is about 18% reduction, which is very significant.

For the fatigue strength under normal stress, the results presented in Chapter 3 can be used directly. Since the steel of the girder falls into fatigue category B (Class B) based on BS 7608 (2014), the ratio (k_r) of fatigue strength to ultimate strength can be taken as $k_r = 0.31$ (BS 7608 2014). Since the ultimate strength reduces due to corrosion, so does the fatigue strength with the constant ratio k for the given fatigue category. From Equation (6.9) (from the same immersion solution), it can be obtained that the residual ultimate strength is $\sigma_u = 282$ for $\sigma_{u0} = 430$ MPa. Therefore, the remaining fatigue strength is $\sigma_f = k_r \sigma_u = 0.31 \times 282 = 87.42$ MPa. This means that for a load cycle of 10^7, the maximum tensile strength in the steel reduces to 87.42 MPa.

The final step of the procedure, i.e., Step (9), is to assess the structural capacity of the beam which should follow a code or standard for required criteria. Let the strength criterion be flexural capacity. For the simply supported beam, the maximum bending moment occurs at mid span, which fluctuates due to cyclic live load. According to basic structural mechanics, the bending moment M at the mid-span can be determined as follows:

$$M = \frac{wl^2}{8} \tag{6.12}$$

where w is the uniformly distributed load in kN/m, and l is the span in m. Assuming that the load factors for dead and live loads are 1.25 and 1.5, respectively, from Equation (6.12), the maximum bending moment can be determined as 180 kN/m. The maximum flexural stress produced by this bending moment can be determined as follows:

$$\sigma = \frac{M}{Z} \tag{6.13}$$

where Z is the effective section module of the beam, which can be calculated for a rectangular hollow section before and after corrosion as 2.24×10^6 and $2.17 \times 10^6 \text{mm}^3$, respectively. After corrosion, the section wall is reduced by 0.3 mm perimetrically. Therefore, the maximum flexural stress in the streel member is 80.2 and 82.8 MPa before and after corrosion, respectively.

For the assessment of flexural capacity, the criterion is adopted from a design code, which is load and resistance actor design as follows:

$$\varnothing R \geq \Sigma \gamma_i S_i \tag{6.14}$$

where $\varnothing$ is the resistance factor to be determined from a design code, R is the resistance, i.e., flexural strength in this example, γ_i is the ith load factor to be determined from a code, i.e., 1.25 and 1.5 for dead and live loads as used above, S_i is the ith load effect, i.e., flexural stress in this example, and Σ denotes load combination, which in this example is the combination of dead and live loads. Assuming that the resistance factor for flexural capacity is 0.9 (such as AS4100), it can be obtained from Equation (6.14) that the remaining flexural strength (resistance) of the girder is $0.9 \times 253 = 228$ MPa, which is greater than the maximum stress (load effect) of 82.8 MPa in the beam. Thus, the beam is safe for bending. It should be noted that the max stress in the beam before corrosion is 80.2 MPa, which is only about 3% increase due to reduction of wall thickness of the section. Compared with reduction of yield strength from 310 to 253, which is 18%, consideration of sectional reduction can grossly overestimate the capacity of the corroded steel structural members.

For the fatigue strength, assuming that the resistance factor for fatigue capacity is 0.8, it can be obtained from Equation (6.14) that the remaining fatigue strength (resistance) is $0.8 \times 87.42 = 69.94$ MPa, which is smaller than the maximum stress range (load effect) of 82.8 MPa. Thus, the beam is unsafe for fatigue under the cyclic loading with a magnitude of 2.5 kN/m and the number of load cycle 10^7.

It needs to be noted that more sophisticated methods are available for structural assessment including those from regulatory codes and standards, and those developed from research such as time-dependent reliability methods, which is beyond the scope of this chapter but is worth noting. Details of such methods can be referred to in publications of the authors, such as Li (2004), Yang et al. (2018) and Furozi et al. (2018), to name a few.

6.3.3 Cast iron structures

Cast iron is mostly used for underground pipelines. Thus, cast iron pipes are taken as examples for design and assessment of corrosion-prone pipes buried in soil. Again, the purpose of these examples is to illustrate how to apply test data presented in previous chapters to practical design and assessment of cast iron pipes subjected to corrosion. In these examples, it is assumed that only corrosion-related changes are considered in the design and assessment of cast iron pipes.

Example 6.6: Design

Consider a cast iron pipe to be laid in soil which is required to be designed for tensile strength as ultimate limit state. Since the pipe is buried in soil, a sacrificial loss of its wall thickness of 0.1% per year is accounted for in the design due to expected corrosion. To make use of test data on steel corrosion and its effect on degradation of tensile strength as presented in Chapter 4, an acceleration factor needs to be determined as per Steps (3) and (4) of the general procedure outlined in Section 6.3.1. Since the pipe is buried in acidic soil, test data produced from the simulated soil solution with pH = 5.5 can be the closest to this service environment. From Figure 4.4, the corrosion loss increases almost linearly with immersion time for cast iron in such environment. From a simple regression analysis, it can be obtained that

$$C = 0.0002t \tag{6.15}$$

where C is the corrosion loss in mm, and t is immersion or accelerated time in days. The coefficient 0.0002 is the corrosion rate in the simulated soil solution in mm per immersion or accelerated day. The sacrificial corrosion rate in the service environment is 0.1% per natural year. For a pipe with a wall thickness of 10 mm, 0.1% per year is 0.01 mm/year, i.e., 0.00003 per day. From these two sets of data, the acceleration factor can be determined as 6.7 (= 0.0002/0.00003). This acceleration factor can be used to gauge the accelerated time to real natural time as an estimation.

From Figure 4.10, the relative reduction of tensile strength can be approximated by a linear relation (for simple illustration) as follows:

$$\sigma_t(t) = \sigma_{t0}\left(1 - \frac{0.04t}{100}\right) \tag{6.16a}$$

where σ_{t0} and σ_t are the original tensile strength and the tensile strength at time t in accelerated days, respectively. For a given acceptable reduction of 10%, such as resistance factor of 0.9 as specified in some codes and standards, from Equation (6.16a)

$$\frac{\sigma_t(t)}{\sigma_{t0}} = \left(1 - \frac{0.04t}{100}\right) = 0.1 \tag{6.16b}$$

it can be obtained that t is 2,250 accelerated days. Using the acceleration factor of 6.7, it is equivalent to $41\left(= \frac{22{,}250 \times 6.7}{365}\right)$ natural years. Since the tensile strength is the design criteria, the design service life for this cast iron pipe can be estimated as 41 years.

On the other hand, for a given design service life of 100 years, the sacrificial loss of the wall thickness can only be designed for about 0.005 mm/year or 0.05% for the wall thickness of 10 mm. This calculation is straightforward, the details of which are omitted here. It should be noted that this example is purely for illustration of how to apply test data presented in this book to design of corrosion-affected or corrosion-prone structures. There are many assumptions acquiesced in this example, such as linear corrosion and linear reduction of tensile strength. As discussed in previous chapters, this is only technical and not methodological, which is the purpose of this example.

Example 6.7: Assessment

If it is recognised that assessment of corrosion-affected bridge structures is very complicated, it is even more complicated for buried pipes simply because there is a soil and pipe interaction, and most of failures of corroded pipes are by fracture. To assess fracture failure of pipes, sophisticated methods and computing tools are necessary, which is way beyond the scope of the book. Since the purpose of this chapter is to demonstrate the application of test data presented in previous chapters, a simple and commonly used assessment criterion is adopted which is strength. For pipes, a unique strength of the pipe is modulus of rupture which has not been demonstrated in Section 6.3.2 and hence is added in the example.

Now consider a cast iron pipe buried in soil for 69 years. Due to aging and long-term exposure to corrosive soil, the structural integrity of the pipe is of concern. The pipe is subjected to an external point load (treated as dead load) that produces maximum stress (load effect) of 210 MPa in the pipe. In this example and for simplicity, it is assumed that all parameters other than corrosion in calculating the stress are constant. Following the procedure outlined in Section 6.3.1, it is assumed that Steps (1)–(3) have been undertaken as described above. From Step (4), a visual inspection needs to be carried out to know the corrosion damage to the pipe as measured by corrosion loss. Site inspection shows that the corrosion is uniform. The average corrosion loss in thickness as measured in situ after cleaning the rusts is 0.05 mm. Since the pipe is still in service, it is not possible to determine its corrosion rate, i.e., actual corrosion loss over time. It is assumed that corrosion initiated in the beginning of its service.

The concept of time transformation is used to gauge the actual environment to an accelerated environment as per Step (6) of the general procedure. Using data from immersion tests in simulated soil solution with pH = 5.5 (the closest to marine environment from all tests data), 0.05 mm in 69 years in natural environment is equivalent to 167 days in accelerated time, i.e., immersion time in the soil solution, as determined from Equation (6.15). With this equivalence, the acceleration factor is 151 (= 69 × 365/167), as determined from Equation (6.3). It should be noted that, strictly speaking, the acceleration factor should be determined from the change rate of the measured parameters, but in this example, it is not possible to estimate the change rate of corrosion, i.e., corrosion rate, accurately since there is only one data point on corrosion loss.

With the equivalent time in immersion tests, the residual strength or remaining strength of the corroded cast iron can be determined accordingly from the test results presented in Chapter 4. Following Equation (6.9), the relationship between the relative reduction of tensile strength and immersion time in the solution with pH = 5.5 can be obtained from Figure 4.10 of Chapter 4 through regression analysis, which is expressed as follows:

$$\sigma_t(t) = \sigma_{t0}\left(1 - \frac{-0.0002t^2 + 0.102t}{100}\right) \tag{6.17}$$

where σ_{t0} and σ_t are the original tensile strength and tensile strength at time t in immersion or accelerated days, respectively. The coefficient of determination for Equation (6.17) is almost 1. From Equation (6.17) and with $t = 167$ accelerated days (69 natural years) and $\sigma_{t0} = 320$ MPa, the residual tensile strength of cast iron can be determined as $\sigma_t = 283$ MPa. This is about 11.5% reduction, which is very significant.

Using the load and resistance design criterion for assessment, as widely adopted in design codes, with a load factor of 1.25 (underground load is treated as dead load) and a resistance factor of 0.9, it can be obtained from Equation (6.14) that

Tensile capacity = 0.9 × 283 = 255 MPa

Tensile stress = 1.25 × 210 = 263 MPa

Since the tensile capacity (resistance or strength) is less than the tensile stress (load effect), the pipe failed after 69 years of service due to corrosion. It is very clear from the calculation that if the corrosion-induced degradation of tensile strength is not considered, the pipe would have been assessed to be safe, that is,

$$0.9 \times 320 = 288 > 1.25 \times 210 = 263 \text{ MPa}$$

This is very significant practically when corrosion-affected cast iron pipes are assessed.

Modulus of rupture is an important strength of pipes, the residual modulus of which needs to be assessed as well. The procedure can be the same as that for tensile strength assessment. However, since there is test data on reduction of modulus of rupture from natural soil, it can be applied directly. From Figure 4.13b, a relationship between the relative reduction of modulus of rupture of a ring section and immersion time in the solution with pH = 5.5 can be obtained from regression analysis, which is expressed as follows:

$$\sigma_m(t) = \sigma_{m0}\left(1 - \frac{0.41t}{100}\right) \tag{6.18}$$

where σ_{m0} and σ_m are the original modulus of rupture and modulus of rupture at time t in natural years, respectively. The coefficient of determination for Equation (6.18) is 0.94. For $\sigma_{m0} = 350$ MPa and at 69 years, the residual modulus of rupture is 251 MPa. This is 28%, which is a very significant reduction.

Again, using the load and resistance design criterion for assessment with a load factor of 1.25 and a resistance factor of 0.9, it can be obtained from Equation (6.14) that

$$\text{Modulus of rupture} = 0.9 \times 251 = 226 \text{ MPa}$$

$$\text{Tensile stress} = 1.25 \times 210 = 263 \text{ MPa}$$

Since the modulus of rupture (resistance) is less than tensile stress (load effect), the pipe failed after 69 years of service. Again, if the corrosion-induced degradation of tensile strength is not considered, the pipe would have been assessed to be safe, that is,

$$0.9 \times 350 = 315 > 1.25 \times 210 = 263 \text{ MPa}$$

Therefore, it is imperative to consider the corrosion-induced degradation of mechanical properties of cast iron when corrosion-affected cast iron pipes are assessed. This is essential from the practical perspective because accurate assessment can prevent unnecessary failures of corrosion-affected structures and possibly casualties.

6.4 SIMULTANEOUS CORROSION AND SERVICE LOADS

One of the potential issues for assessment of corrosion damages to steel and steel structures is that steel is subjected to corrosion and applied load or stress simultaneously. It has been shown in Section 5.5 that the applied stress affects the corrosion behaviour to a certain extent. For example, corrosion loss is up to 70% higher for specimens in acidic solution with stress than those without stress. Also, the reduction of tensile strength is up to 9% higher for specimens in acidic solution with stress than those without stress. This clearly indicates that there is an interaction between corrosion and stress. From both electrochemical and mechanistic points of view, the interaction between corrosion and stress makes sense as demonstrated in Section 5.3. It follows that corrosion tests should be conducted in the presence of stress. This idea was in fact first proposed in 2000 by Li (2000) for corrosion of reinforcing steel in concrete with considerable tests undertaken to illustrate the idea (see Li 2000, 2001 and 2002 for details). For steel structures, it is more imperative to consider the interaction of corrosion and stress, but thus far, most test data are not under the condition of simultaneous corrosion and stress. This section will shed some light on how to conduct corrosion tests in the presence of stress. It is acknowledged that there is always a gap between idea or theory and practice. It is argued that idea is always the first step towards filling the gap.

6.4.1 Testing methodology

A general testing methodology is presented here to serve a few purposes. One is that there is a method to follow that has considered the interaction of corrosion and service load. The second purpose is that from this method, improvement can be made to be more rational and practically operational. The third is that, in lieu of an accepted method, the proposed method can serve as a guide for conducting corrosion tests on steel and steel structures or members.

To create an environment in which both corrosion and stress can occur simultaneously, an environmental chamber is needed, which can be small using coupons as test specimens or large using a structural member as a test specimen. In analogy to the testing methodology first proposed by Li in 2000 (Li 2000), a large environmental chamber can be built to accommodate the specimens and, in the meantime, create an environment to which the specimens are meant to exposed. A sketch of a large such environmental chamber is shown in Figure 6.1.

The chamber itself should be built with solid materials, and in particular, the joints must be airtight. The inner surface of the chamber should be lined with corrosion resisting materials, such as chromium alloy stainless steel. The chamber should be equipped with facilities for controls of basic climate parameters, such as temperature, relative humility and aggressive agents, such as chloride concentration. Inside the chamber, various environments can be simulated. For example, a water spray system can be installed to simulate marine and coastal environments where the tides and splashes can be simulated. Also, the water can be in the form of different solutions,

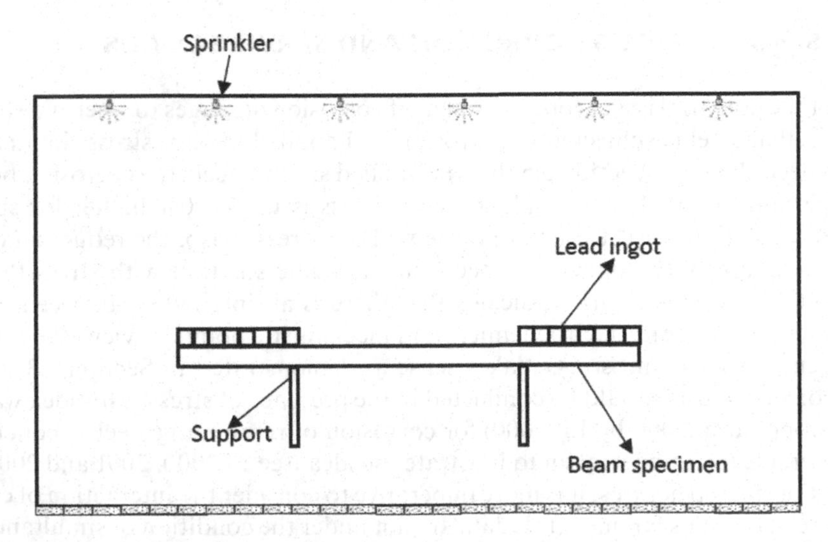

Figure 6.1 A sketch of the proposed environmental chamber (elevation).

such as NaCl solution to simulate sea water or an HCl solution to simulate acidic solutions. In the meantime, various loadings can be applied to the specimens to achieve simultaneous corrosion and stress.

Usually, one environmental chamber is purposely designed primarily to simulate one or two types of environments. It is also possible that the chamber is designed comprehensively for multiple purposes by either equipping it with sophisticated facilities or partitioning the space to simulate different environments for different purposes. Of course, the cost for the latter would be also very high. Depending on the materials used, in particular the internal lining materials, such as chromium alloy stainless steel, the design life for the environmental chamber is usually less than 10 years. For example, a large environment chamber, which is believed to be the first one in such a scale and for such purpose, was built at Monash University Australia in 1998 for a design service life of 7 years (Li 2000).

Test specimens are constructed on a site outside the chamber and moved inside the chamber to be exposed to the designated environment. The service load can be applied to the specimens either manually by weights, such as led ingots as schematically shown in Figure 6.1, or by hydraulic loading system as long as the space permits. The set-up of the test is the same as that in normal laboratories, but the instrumentation, including loading, needs more care due to the aggressive environment. Experience of testing in a harsh environment suggests that manual measurement is preferred as much as possible and wherever it is possible. Also, the measurement under simultaneous corrosion and stress cannot be as comprehensive as normal structural testing where complicated electrical devices cannot be used for the duration of the tests due to the aggressive environment. The main purpose is observation and macromeasurement manually as much as possible.

Most measurements, either for corrosion or for mechanical properties, can be conducted by removing the test specimens out of the chamber in the same way as removing specimens out of the simulated solution. Thus, there should be a sufficient number of specimens placed in the chamber exposed to the environments and stress. At a certain point in time, one or more specimens are taken out of the chamber to carry out corrosion measurement, such as corrosion loss, and mechanical testing, such as tensile strength. A minimum three of such time points should be allowed for to create a relationship of test measurement as a function of time. It may be noted that data produced from such tests are only representative of the environmental conditions simulated in the chamber. To make use of such data, a calibration process is required, which has been covered in Section 6.2.

In principle, the testing methodology presented above can be applied to all simulated tests under combined corrosion and loading conditions. In practice, however, specific conditions require specific design of the environmental chamber and the test up in the chamber. Two examples of tests under simultaneous corrosion and corrosion are provided below to elaborate further.

6.4.2 Combined corrosion and bending

Commonly encountered loads in practical structures are tension, compression, flexure and shear, all of which can be simulated in a bending test as long as the size of the

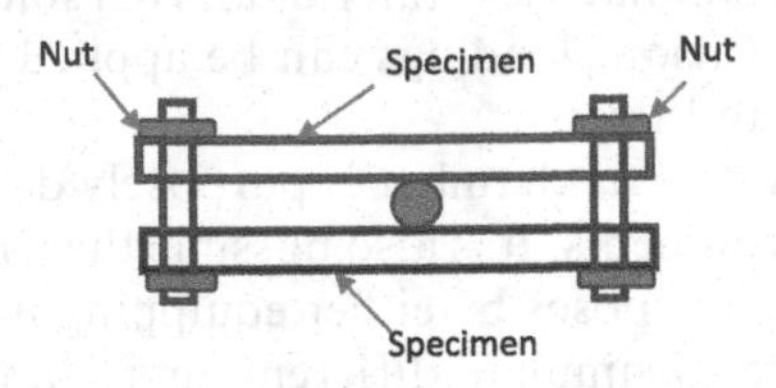

Figure 6.2 A sketch of coupling a pair of specimens.

environmental chamber is sufficiently large to house such specimens for testing in the designated environment. For material tests using a coupon under simple loading conditions, such as tension, a container can act as an environmental chamber as presented in Section 5.5.1. Even for a little more complex loading condition, such as bending, it is still possible to use a container for simultaneous corrosion and bending tests, as schematically shown in Figure 6.2. Two specimens can be coupled together where loading is applied at each end by tightening the nuts to a designated level of stress. Then the test rig is placed in a container filled with acidic solutions as those tests without stress presented in Chapters 3 and 4. The advantage of this coupling set-up is that the specimen is under multiple states of stress, such as tension, compression and shearing. Together with simultaneous corrosion, this test can investigate the effect of different stresses (other than normal only) and/or a combination of stress and corrosion and their interactions. After immersion in the solution for a certain period of time, the rest of the test procedure would be the same as that without stress, such as taking the specimens out of container, cleaning, corrosion measurement and mechanical testing. Details of this can be referred back in previous chapters.

For application to steel structure, it is desirable to have reasonably large size of specimens representing real structural members. This poses difficulties in two folds: one is that a large environmental chamber needs to be built to accommodate a reasonable size of structural member. For example, a simplest structure can be a simply supported beam, and a reasonable size can be 3 m in length. Thus, the size of the chamber can be 6 (length) × 3 (width) × 3 (height) m for comfortable use and health and safety requirement. The other difficulty is loading. Since the size of the specimen is large, a reasonable amount of loading is required to produce significant stress in the specimens. To achieve this, it is better to make use of cantilever configuration simply because it creates large stress in the specimens and multiple stress states, such as tension, compression and shearing, as shown in Figure 6.1.

Loading is always the difficulty in tests with simultaneous corrosion and stress in particular, when the size of specimens is large. Experience in this kind of tests suggests that dead weight, such as led ingots, is preferred to hydraulic loading because of the aggressive environment of corrosion. For the cantilever beam, the manual loading has more advantage since it can produce larger stress for the given specimen size and structural span. It may be noted that hydraulic loading is not impossible. It is a matter of cost-effectiveness. For example, if the hydraulic loading system is only used for once for the designated test, of course hydraulic loading is more accurate and flexible than manual loading.

Measurement of test variables and parameters during the test is another difficulty frequently encountered in tests with simultaneous corrosion and stress, regardless of the size of specimens. Experience of testing in a corrosive environment suggests that a mechanical device is preferred as much as possible to an electronic device. For example, strain gauges cannot survive for long in a corrosive environment for many reasons, but aggressivity is the main one. Fortunately, from phenomenological and mechanistic points of view, which are the approach adopted in this book, observation is important during the test process and key measurements, such as corrosion loss and mechanical strength, are conducted by removing the test specimens out of the chamber, effectively avoiding the aggressive environment for further testing.

6.4.3 Combined corrosion and fatigue

As both corrosion and fatigue are long-term phenomena, the effect of corrosion on fatigue strength of steel should be investigated in the presence of cyclic loading to account for their interaction. In other words, tests on corrosion effect on fatigue need to be carried out under the simultaneous corrosion and cyclic loading. This is only possible if a steel structure is decommissioned after a certain period of service. Then steel plates can be cut from it for fatigue test as presented in Section 3.4.1. However, realistically, obtaining a decommissioned steel structure poses great challenge. Likewise, conducting fatigue tests under simultaneous corrosion and cyclic loading equally presents great challenges in laboratories and perhaps even greater challenges than to find a decommissioned steel structure. For this purpose, a testing methodology to achieve simultaneous corrosion and fatigue is presented herein.

The rational for this testing methodology is to design a small chamber like a box that can contain the corrosive solution for the specimen to be immersed whilst it is under cyclic loading, known as a chamber system. The chamber system is designed to create an environment of simultaneous corrosion and cyclic loading for specimens. The geometry of the chamber is designed to suit the purpose and large enough to contain the test specimen. The chamber is comprised of a clamp with a holder at the bottom, a container connected with the clamp and a cover as schematically shown in Figure 6.3. The chamber, including connecting bolts, should be made of high-strength corrosion-resistant steel, such as stainless steel, to prevent its failure during the corrosion and fatigue combined tests.

Assembling of a test rig for simultaneous corrosion and fatigue test is as follows (Li et al. 2019). First, the chamber system is coated with Vaseline thoroughly to protect it from corrosion. Then, the test specimen is connected to the bottom of the chamber by four high strength bolts, such as M6 grade 12.9 bolts with a yield strength of 1,080 MPa. The bolts should be preloaded to ensure the tightness. Afterward, corrosive solution, such as HCl solution, as used in Section in 3.1.1, is poured into the container up to about the height to immerse the specimen. The solution in the chamber can be changed periodically to ensure the designated acidity (pH) to be constant. The height of the container is designed such that there is sufficient space left after it is filled with solution to ensure adequate air inside the chamber. The chamber is then covered on the top to prevent solution spilling. There is a small opening on the top cover for the specimen to be outside the container. During the test, any gaps between the specimen

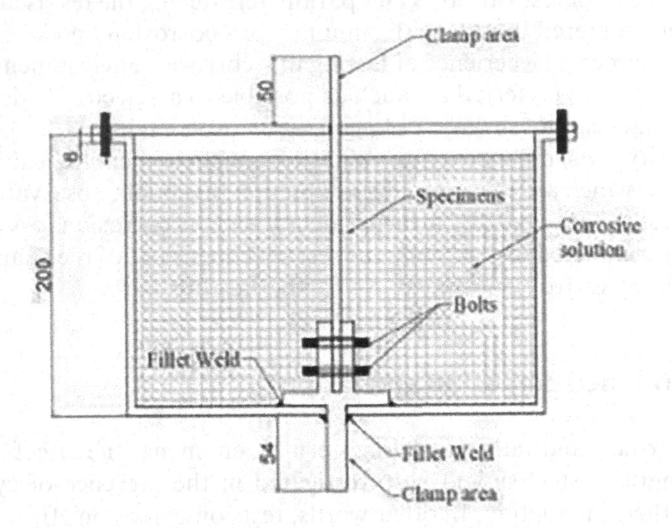

Figure 6.3 Schematic of the chamber system for combined corrosion and fatigue test.

Figure 6.4 Set-up of combined corrosion and fatigue test.

end and the opening is sealed. The cover of the chamber can be lifted from time to time during the test to monitor corrosion. The chamber system is then placed in the fatigue test machine, such as MTS 100, with the bottom end of the chamber clamped to the test machine and the top end of the specimen fixed to the other clamp of the test machine to perform simultaneous corrosion and fatigue test. A photo of such set-up is shown in Figure 6.4. Once set up, axial constant cyclic loading is applied on specimens.

The test rig presented here has a clear advantage over previous tests since it creates a simultaneous corrosion and fatigue environment, which simulates the real service conditions in engineering practice.

6.5 NANOMECHANICS OF CORROSION

By now, sufficient evidence has been presented in the book that corrosion does cause degradation of mechanical properties of corroded steel or ferrous metals in general. This not only includes test data produced from the laboratory where the corrosion is simulated, which is acknowledged somehow artificial, but importantly also includes test data on steel collected from steel structures after service for different periods of time, which effectively corroborate the test data from the laboratory. Together with reoccurrence of unexpected collapses of corroded steel structures, and importantly, increased knowledge of corrosion, in particular its diagnosis and prevention, it is quite convincing that the test data presented in the book provide at least part of solution to the problem of failures of corroded structures: the corrosion causes degradation of mechanical properties of corroded steel, and this degradation should be considered in the design and assessment of steel and steel structures affected by corrosion. Not doing so may result in continual unexpected failures of corroded steel and steel structures.

Mechanisms of degradation of mechanical properties of corroded steel are also discussed but only post-mortemly. Relationships or models between corrosion and reduction of mechanical properties are also developed but only empirically. There is still a lack of fundamental explanation and understanding on why and how corrosion affects the mechanical properties of corroded steel. It is believed that there is a long way to go to clearly understand the fundamental mechanism of degradation of mechanical properties of corroded steel. This chapter intends to move a step towards a clear understanding of why the mechanical properties degrade due to corrosion by presenting an idea to explore the problem from the root bottom of atoms of steel.

6.5.1 Basic idea

The reoccurrence of collapses of corroded steel structures suggests that existing research on corrosion has missed a critical point – the mechanism for degradation of mechanical properties of corroded steel must be accurately determined. With this knowledge, structural collapse due to corroded steel can be accurately predicted as it is the mechanical properties of steel material that provide strength to the structure to prevent its collapse. This is the major gap in knowledge that hinders further advancement of corrosion science and hampers it application to accurate prediction of collapse of corroded steel structures. The root of the problem can be how the strength of corroded steel degrades from its constituent sources. It is hypothesised that the changes of atomic lattice due to corrosion are the root causes for degradation of mechanical properties of corroded steel. This chapter intends to track deep down to the bottom of steel constituents, i.e., atomic lattice, to reveal this root cause and determine its effect on degradation of mechanical properties of corroded steel.

On the atomic scale, the corrosion of steel can be understood as severe oxidation of steel surface atoms in contact with air and moisture. The physical and mechanical

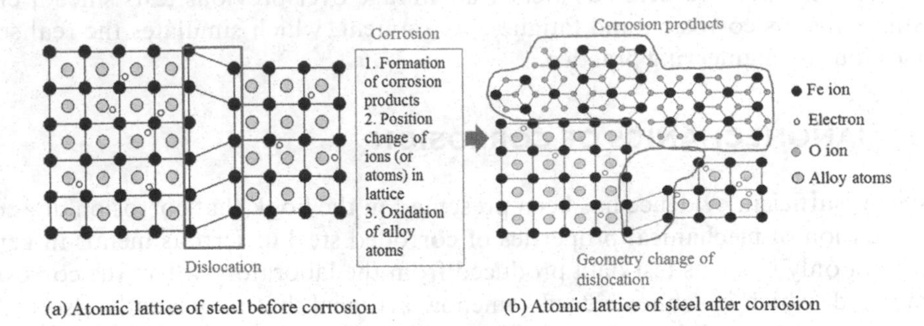

Figure 6.5 Atomic lattice of steel before and after corrosion.

relationship between atoms can be dealt with by quantum mechanics, which treats the steel as a perfect iron (Fe) atomic lattice on the atomic scale. Since steel is always used in structures, the initial atomic lattice of steel is not perfect but includes defects, such as dislocations and impurities, such as alloy atoms, as shown in Figure 6.5 schematically (MacDonald 1999). When steel corrodes, the atomic lattice changes accordingly as shown in Figure 6.5b due to reaction with oxygen and water (MacDonald 1999, Revie and Uhlig 2009, Marcus and Maurice 2017). If the changes of atomic lattice can be determined quantitatively under certain environmental conditions, such as air (oxygen), moisture (water) and temperature, it is possible to determine the mechanical properties of steel at the atomic level. Clearly, it is fundamental to examine quantitatively how corrosion changes the atomic lattice, such as arrangement and distribution of atoms (ions) and electrons, dislocations and impurities, as shown in Figure 6.5b, and importantly, how these changes alter the properties of the atomic lattice, such as interatomic potential and mechanical strength, from which the ultimate mechanical strength of corroded steel can be deduced.

It is understood from quantum mechanics that the arrangement of ions and distribution of electrons (Figure 6.5) are related to the interatomic potential of the atomic lattice, which represents the binding energy required to separate ions (or atoms) from their equilibrium spacing in the lattice to an infinite distance (Di Tommaso et al. 2012). This relation can be expressed as follows:

$$u_{\mathrm{IP}} = \frac{1}{2}\sum_{i=1}^{N}\sum_{j\neq i}^{N} u\left(r_{ij}\right) + \sum_{i=1}^{N} E\left(\rho_i\right) \tag{6.19}$$

where u_{IP} is the interatomic potential, N is the total number of ions and atoms within the given steel, r_{ij} is the distance between two ions or atoms (such as ions i and j), $u()$ is the pair potential between two ions (or atoms), $E()$ is the embedding function that determines the energy required to place ion (or atom) i in the background electron density, and ρ_i is the host electron density at ion (or atom) i due to the remaining atoms of the system. Equation (6.19) has been used to predict interatomic potentials for steel with a regular (known) atomic lattice arrangement, such as Figure 6.5a (Simonelli et al. 1992) but how to consider the changes of atomic lattice, such as r_{ij}, $u()$, $E()$ and ρ_i due to corrosion, such as Figure 6.5b remains a big challenge.

Through molecular dynamics simulation, Meyers and Chawla (2008) have established the relationship between interatomic potential and mechanical strength of an intact atomic lattice, i.e., an atomic lattice of pure iron crystal. However, the atomic lattice of steel material always contains dislocations and impurities which also change after corrosion (see Figure 6.5a and b). A key to further advance the corrosion science from a mechanistic perspective is to establish the relationship between interatomic potential and mechanical strength of atomic lattice of steel before and after corrosion. Also, the mechanical strength of the intact atomic lattice cannot represent the strength of steel material because both dislocations and impurities can change the attractive and repulsive forces between ions and alter the interatomic potential of atomic lattice (Buehler and Gao 2006, Li et al. 2020d). Furthermore, corrosion will further change the interatomic potential of atomic lattice. No theory or research exists to explain and determine the relationship of mechanical strength between the atomic lattice and steel before and after corrosion. Therefore, the basic idea proposed here is to

1. quantify the changes in atomic lattice
2. determine energy-strength relation of atomic lattice
3. correlate strength of atomic lattice and that of steel.

6.5.2 Mapping of atomic lattice

Experiments are proposed to map out the topography of atomic lattice due to corrosion. Tests will be conducted using a number of state-of-the-art facilities, including micro-computed tomography (micro-CT) and a scanning transmission electron microscope equipped with an electron energy loss spectroscopy system (STEM-EELS) in the laboratories with microscopy and microanalysis facility. Specimens of $2 \times 4 \times 1$ mm will be obtained from corroded steel. Corrosion losses of these specimens will be measured, and the results are classified into three different corrosive conditions (mild, moderate and severe). Three specimens will be used under each condition (plus uncorroded specimens as benchmark) for testing. Micro-CT will create the 3D topography of the corroded surface of each specimen at different scanning locations. There will be eight scanning locations on each specimen with a scan area of $50 \times 50\,\mu m$; this area is large enough to include both bulk grain, which contains an intact lattice, and grain boundaries, which contain different defects and alloys. Nanoscale samples will be cut at each scanning location with different cutting angles (0, 45°, 90°) by a focused ion beam microscope, as shown in Figure 6.6 to measure surface modification and functional structure of the atomic lattices. Different cutting angles will be used so that the topography of the atomic lattice can be mapped out in 3D. Nanoscale samples will have the shape of flat tensile test specimens with a thickness of 50 nm and a tested area of $3 \times 3\,\mu m$. STEM will produce images of the atomic lattice at each tested area. Different types of ions (or atoms) on images can then be classified through EELS mapping by assigning different colours to them as shown in Figure 6.7. Also, synchrotron X-ray photoelectron spectroscopy will be used to characterise the elemental composition and chemical/electronic state of elements on the corroded surface of each specimen, from which the oxidation state of irons can be monitored. In mapping the changes of atomic lattice, the critical depth that activates these changes can be determined. This

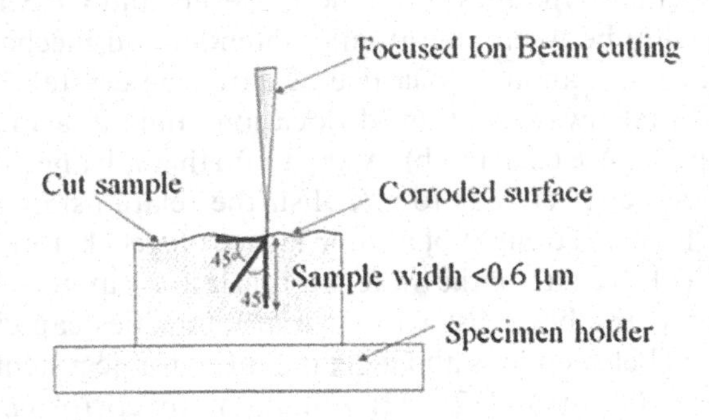

Figure 6.6 Configuration of sample preparation.

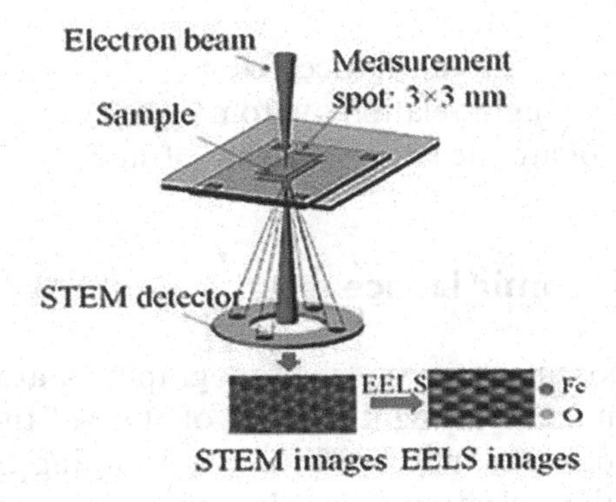

Figure 6.7 Schematic of STEM measurement.

is very significant both theoretically and practically because if the corrosion loss is less than this critical depth, it would not cause changes of atomic lattice and hence no degradation of mechanical strength or properties of the corroded steel.

After STEM-EELS analysis, massive data will have been produced from which analytics should be developed to quantify the topography of the atomic lattice. A deep learning algorithm can be employed to determine the 3D coordinates of the topography of the atomic lattice after corrosion as shown in Figure 6.5b. The algorithm will be designed based on a fully conventional network (FCN) developed in the Python programming language, which contains multiple hidden layers as shown in Figure 6.8. The Laplacian of Gaussian (LoG) blob detection method will be applied to the output of a FCN to detect features in digital images (such as green dots in Figure 6.8) and determine their coordinates. Software for material science, such as VASP, LAMMPS and GATAN, is available and can be employed for the analysis.

In this algorithm, training will be first provided with the following steps: (i) a small portion of EELS and STEM images produced from above will be fed into a FCN (a deep learning algorithm commonly applied to analyse visual imagery); (ii) classification criteria will be set, which can be the assigned colour of each ion for EELS images or the distance between each ion (or atom) for STEM images (if the distance differs from the lattice constant of iron, it can be dislocations as in Figure 6.3); (iii) FCN will

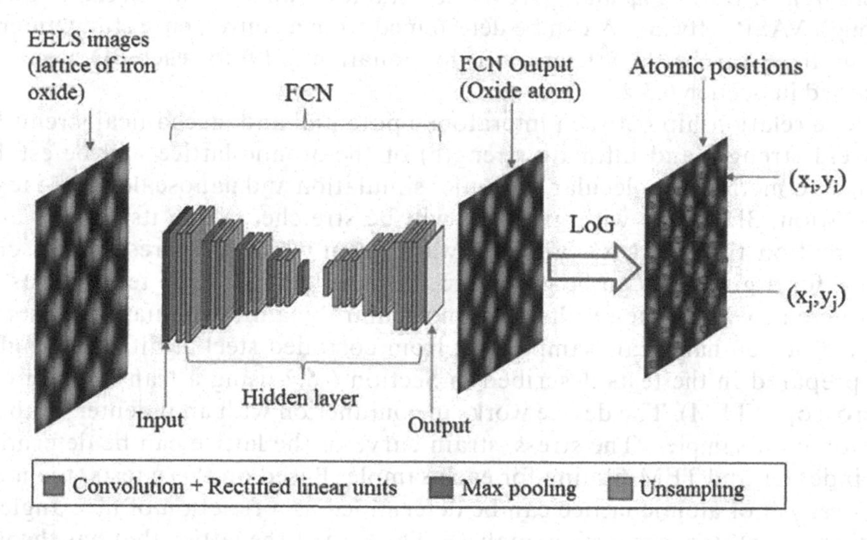

Figure 6.8 Configuration of FCN and LoG.

produce a binary image with small blobs, which shows ions (or atoms) in a specific classification; (iv) the coordinates of these blobs will be determined from the LoG blob detection method; and (v) coordinates will be compared with those determined from GATAN software to optimise the algorithm. After the training, the algorithm will be applied to EELS and STEM images to classify different ions (or atoms) and dislocations and determine their coordinates. Also, by combining the ionic coordinates of samples at different cutting angles, the layout of atomic lattice can be mapped in 3D at each scanning location. The electron density will also be determined by density functional theory based on lattice layout through VASP software. From the data produced here, models on changes in atomic lattice due to corrosion, as measured by corrosion loss or depth, can thus be established.

6.5.3 Model development

With the coordinates of ions and atoms mapped out in Section 6.5.2, the distance between each ion (or atom), r_{ij}, and the host electron density, ρ_i, can be then determined from analytic geometry and density functional theory, respectively. The pair potential function $u(r_{ij})$ and embedding function $E(\rho_i)$ can be determined as follows (Belonoshko and Ahuja 1997):

$$u(r_{ij}) = \left(\frac{A}{r_{ij}}\right)^n + D\left[e^{-2\alpha(r_{ij}-r_0)} - 2e^{-\alpha(r_{ij}-r_0)}\right] - \frac{C_{vdW}}{r^6_{ij}} \tag{6.20}$$

$$E(\rho_i) = C \sum_{j(j \neq i)}^{N} \rho_j^{\frac{1}{2}} \tag{6.21}$$

where $A, D, n, \alpha, r_o, C_{vdW}$ and C are the adjustable parameters, which can be quantified through VASP software. N can be determined from a convergence study. Interatomic potential u_{IP} can be then determined by Equation (6.19) for each 3D atomic lattice obtained in Section 6.3.2.

The relationship between interatomic potential and mechanical strength (such as yield strength and ultimate strength) of the atomic lattice will be established using two methods: molecular dynamics simulation and nanoscale tensile tests. For simulation, 3D lattice with given u_{IP} will be stretched along its axis by applying deformation through LAMMPS software, from which the stress–strain curve of lattice for a given u_{IP} can be generated. In parallel, nanoscale tensile tests will be conducted to verify the results from molecular dynamics simulation. Tests will be carried out on nanoscale samples cut from corroded steel at different angles that are prepared in the tests described in Section 6.5.2 using a transmission electron microscope (TEM). The device works in conjunction with an indenter to apply tensile force on samples. The stress–strain curve of the lattice can be determined by the indenter and TEM filming for each sample. Based on these tests, the mechanical strength of atomic lattice can be determined as a function of face angle θ (i.e., cutting angle) from regression analysis. The face of the lattice that has the weakest tensile strength will be the tensile strength of the 3D atomic lattice for a specific interatomic potential u_{IP}, which will be compared with molecular dynamics simulation for verification.

A similarity method will be developed to correlate the stress–strain relationships of atomic lattice and steel after corrosion. Macroscale steel specimens with different corrosive conditions as specified in the tests described in Section 6.5.2 will be tested to determine their mechanical strength, following ASTM E8/E8M-16a. Through LAMMPS software, the stress–strain curves of atomic lattice will have been developed, from which the curve that has the lowest stress at a specific strain will be compared with the average stress–strain curve of specimens from microstructural test by similarity measurements. The similarity between the stress–strain curve of atomic lattice and that of steel can be quantified by the unitless objective function as follows (Cao and Lin 2007):

$$f = \frac{1}{N_1}\sum_{i=1}^{N_1}\left[\frac{\sigma_i^s - \sigma^l\left(\varepsilon_{N1}^l \varepsilon_i^s / \varepsilon_{N1}^s\right)}{S_{\sigma,i}}\right]^2 + \left(\frac{\varepsilon_{N1}^s - \varepsilon_{N1}^l}{S}\right)^2 \tag{6.22}$$

where f is the objective function, N_1 is the total number of data points on the stress–strain curve of steel and $N_1 \approx 900$. σ_i^s and ε_i^s are the stress and strain values for the stress–strain curve of steel at data point i. σ_i^l and ε_i^l are the stress and strain values for the stress–strain curve of the atomic lattice that has the shortest distance from $(\sigma_i^s, \varepsilon_i^s)$. In Equation (6.22), $\sigma^l\left(\varepsilon_{N1}^l \varepsilon_i^s / \varepsilon_{N1}^s\right)$ is the stress of atomic lattice at strain $\varepsilon_{N1}^l \varepsilon_i^s / \varepsilon_{N1}^s$. $S_{\sigma,i} = 0.1\sigma_i^s$ and $S = 0.1\varepsilon_{N1}^s$. The stress–strain curve of atomic lattice and that of steel can be correlated by the smallest f-value at each corrosive condition, as the similarity between them increases with the reduction of f. The parameters in Equation (6.22) are also explained in Figure 6.9.

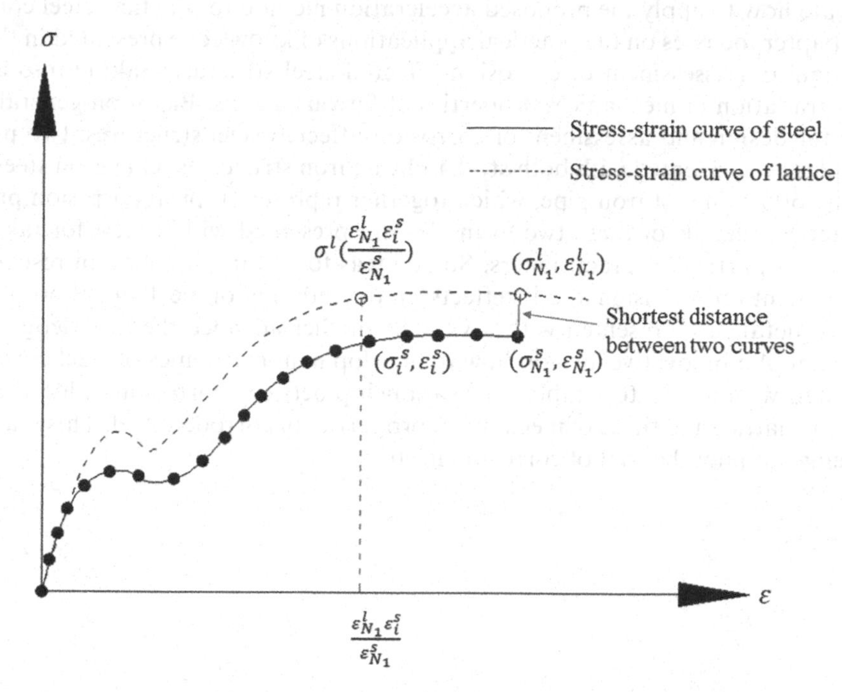

Figure 6.9 Similarity between stress–strain curves of lattice and bulk steel.

The stress–strain curve of lattice that can fit the stress–strain curve of the specimens (f closes to 0) can be determined at each corrosive condition. This stress–strain curve of lattice can represent the stress–strain curve of specimens at the macroscale.

It is acknowledged that what is presented in this section is only an idea. In theory, it is plausible under certain assumptions. In practice, there may be considerable difficulties, and more problems may be encountered. However, the purpose to present this idea is not to show a theory or method but to share a possible direction for future research on corrosion and its effect on degradation of mechanical properties of corroded steel and to further explore the mechanics of degradation of mechanical properties of corroded steel from a more fundamental science perspective rather than phenomenological observations as has already been done in this book.

6.6 SUMMARY

This chapter addresses a fundamental issue of all scientific research, that is, how to apply the knowledge and information produced from the research, as presented in the previous chapters, to practical design and assessment of corrosion-affected steel structures. The basics of similarity theory are presented as a theoretical basis for developing a calibration method for test data produced from the accelerated tests. An acceleration factor method is proposed to achieve the calibration of test data produced from the accelerated environment to that of application. Three examples are presented to illustrate how to apply the proposed acceleration method to practical steel corrosion. The chapter focusses on the practical applications of knowledge presented in the book to design and assessment of corrosion-affected steel structures taking into account the degradation of mechanical properties of ferrous metals. Based on general procedures for design and assessment of corrosion-affected steel structures, two practical examples are presented with both steel and cast iron structures. One is on steel bridge and the other on cast iron pipe, which together represent typical corrosion problems in practice. Details of these two examples are presented with a view for take up by readers, in particular, practitioners. Some ideas for future direction in research and development on corrosion and its effects on degradation of mechanical properties of ferrous metals are presented with a view to further advance the knowledge of steel corrosion. An innovative idea on how to develop nanomechanics of steel corrosion is presented with a view to establish a relationship between corrosion-induced changes of atomic lattice and those of mechanical properties of corroded steel. This is a real big challenge and not the root of corrosion itself.

Bibliography

Abràmoff, M. D., Magalhães, P. J. and Ram, S. J. (2004), "Image processing with ImageJ", *Biophotonics International*, 11, (7), 36–42.

Adasooriya, N. and Siriwardane, S. (2014), "Remaining fatigue life estimation of corroded bridge members", *Fatigue and Fracture of Engineering Materials and Structures*, 37, 603–622.

Ahammed, M. (1997), "Prediction of remaining strength of corroded pressurised pipelines", *International Journal of Pressure Vessels and Piping*, 71, (3), 213–217.

Ahammed, M. (1998), "Probabilistic estimation of remaining life of a pipeline in the presence of active corrosion defects", *International Journal of Pressure Vessels and Piping*, 75, (4), 321–329.

Ahammed, M. and Melchers, R. E. (1997), "Probabilistic analysis of underground pipelines subject to combined stresses and corrosion", *Engineering Structures*, 19, (12), 988–994.

Ahmmad, M. M. and Sumi, Y. (2010), "Strength and deformability of corroded steel plates under quasi-static tensile load", *Journal of Marine Science and Technology*, 15, (1), 1–15.

Aiken, L. (1991), *Multiple Regression: Testing and Interpreting Interactions*, Newbury Park, CA: Sage Publications.

Alamilla, J. L., Espinosa-Medina, M. A. and Sosa, E. (2009), "Modelling steel corrosion damage in soil environment", *Corrosion Science*, 51, (11), 2628–2638.

Aliofkhazraei, M., Rouhaghdam, A. S. and Hassannejad, H. (2009), "Effect of electrolyte temperature on the nano-carbonitride layer fabricated by surface nanocrystallization and plasma treatment on a γ-TiAl alloy", *Rare Metals*, 28, (5), 454–459.

Almubarak, A., Abuhaimed, W. and Almazrouee, A. (2013), "Corrosion behavior of the stressed sensitized austenitic stainless steels of high nitrogen content in seawater", *International Journal of Electrochemistry*, 2013, 1–7.

American Society of Mechanical Engineers and American National Standards Institute. (1985), *Manual for Determining the Remaining Strength of Corroded Pipelines: A Supplement to ANSI/ASME B31 Code for Pressure Piping*, New York, NY: American Society of Mechanical Engineers.

American Water Works Association. (1977), American national standard for thickness design of cast-iron pipe (ANSI/AWWA C101-67), In *AWWA Standards* (pp. 1–78), Denver, CO: American Water Works Association.

Anderson, T. L. (2017), *Fracture Mechanics: Fundamentals and Applications* (4th ed.), Boca Raton, FL: CRC Press.

Andrade, C. and Alonso, C. (1996), "Corrosion rate monitoring in the laboratory and on-site", *Construction and Building Materials*, 10, (5), 315–328.

Arafin, M. A. and Szpunar, J. A. (2009), "A new understanding of intergranular stress corrosion cracking resistance of pipeline steel through grain boundary character and crystallographic texture studies", *Corrosion Science*, 51, (1), 119–128.

Arioka, K., Yamada, T., Terachi, T. and Staehle, R. W. (2006), "Intergranular stress corrosion cracking behavior of austenitic stainless steels in hydrogenated high-temperature water", *Corrosion*, 62, (1), 74–83.

Aryai, V., Baji, H., Mahmoodian, M. and Li, C. Q. (2020), "Time-dependent finite element reliability assessment of cast-iron water pipes subjected to spatio-temporal correlated corrosion process", *Reliability Engineering and System Safety*, 197, 106802–106811.

ASTM International. (2004), *ASTM G102-89 Calculation of Corrosion Rates and Related Information from Electrochemical Measurements*, West Conshohocken, PA: ASTM.

ASTM International. (2006), *ASTM D2487 Standard Practice for Classification of Soils for Engineering Purposes (Unified Soil Classification System)*, West Conshohocken, PA: ASTM

ASTM International (2010a), *A247-10, Standard Test Method for Evaluating the Microstructure of Graphite in Iron Castings*, West Conshohocken, PA: ASTM

ASTM International. (2010b), *ASTM G162-99 Standard practice for conducting and evaluating laboratory corrosion tests in soils*, West Conshohocken, PA: ASTM

ASTM International. (2011a), *ASTM E3-11 Standard Guide for Preparation of Metallographic Specimens*, West Conshohocken, PA: ASTM.

ASTM International. (2011b), *ASTM F1113-87 Standard Test Method for Electrochemical Measurement of Diffusible Hydrogen in Steels*, West Conshohocken, PA: ASTM.

ASTM International. (2012a), *ASTM E399 Standard Test Method for Linear-Elastic Plane-Strain Fracture Roughness KIC of Metallic Materials*, West Conshohocken, PA: ASTM.

ASTM International. (2012b), *ASTM G31-12a Standard guide for laboratory immersion corrosion testing of metals*, West Conshohocken, PA: ASTM.

ASTM International. (2013a), *ASTM D4972 Standard Test Method for pH of Soils*, West Conshohocken, PA: ASTM.

ASTM International. (2013b), *ASTM E112-13 Standard Test Methods for Determining Average Grain Size*, West Conshohocken, PA: ASTM.

ASTM International. (2013c), ASTM G46-94 Standard Guide for Examination and Evaluation of Pitting Corrosion, In *Annual Book of ASTM Standards* (Vol. 03.02), West Conshohocken, PA. ASTM.

ASTM International. (2013d), *ASTM E2809-13 Standard Guide for Using Scanning Electron Microscopy/X-Ray Spectrometry in Forensic Paint Examinations*, West Conshohocken, PA: ASTM

ASTM International. (2014a), *ASTM G4-01 Standard Guide for Conducting Corrosion Tests in Field Applications*, West Conshohocken, PA: ASTM.

ASTM International. (2014b), ASTM G59-97 Standard test method for conducting potentiodynamic polarization resistance measurements, In *Annual Book of ASTM Standards* (Vol. 03.02), West Conshohocken, PA: ASTM.

ASTM International. (2015), *ASTM E466-15 Standard Practice for Conducting Force Controlled Constant Amplitude Axial Fatigue Tests of Metallic Materials*, West Conshohocken, PA: ASTM.

ASTM International. (2016), *ASTM E08/E08M-16a Standard Test Methods for Tension Testing of Metallic Materials*, West Conshohocken, PA: ASTM.

ASTM International. (2017a), *ASTM F1113-87 Standard Test Method for Electrochemical Measurement of Diffusible Hydrogen in Steels (Barnacle Electrode)*, West Conshohocken, PA: ASTM.

ASTM International. (2017b), *ASTM D-2488 Standard Practice for Description and Identification of Soils*, West Conshohocken, PA: ASTM.

ASTM International. (2017c), *ASTM G1-03 Standard Practice for Preparing, Cleaning, and Evaluating Corrosion Test Specimens*, West Conshohocken, PA: ASTM.

ASTM International. (2018a), *ASTM G148-97 Standard Practice for Evaluation of Hydrogen Uptake, Permeation, and Transport in Metals by an Electrochemical Technique*, West Conshohocken, PA: ASTM.

ASTM International. (2018b), *E1820-18 Standard test method for measurement of fracture toughness*, West Conshohocken, PA: ASTM.

ASTM International and NACE International. (2012), ASTM G31-12a Standard guide for laboratory immersion corrosion testing of metals, In *Annual Book of ASTM Standards* (Vol. 03.02). West Conshohocken, PA: ASTM.

Australian Standard. (2007a), *AS 1830 Cast Iron*, Canberra: Australian Standard.

Australian Standard. (2007b), *AS 4100 Steel Structures*, Canberra: Australian Standard.

Australian Standard. (2007c), *AS 1831 Ductile cast iron*, Canberra: Australian Standard.

Australian Standard. (2008a), *AS 3597 Structural and Pressure Vessel Steel - Quenched and Tempered Plate*, Canberra, Australian Standard.

Australian Standard. (2008b), *AS 4312 Atmospheric Corrosivity Zones in Australia*, Canberra: Australian Standard.

Australian Standard. (2016), *AS3679.1 Structural Steel - Hot-Rolled Bars and Sections* Canberra: Australian Standard

Australian Standard. (2017), *AS 5100.6 Bridge Design, Part 6: Steel and Composite Construction*. Canberra: Australian Standard.

Ávila, J. A., Lima, V., Ruchert, C. O., Mei, P. R. and Ramirez, A. J. (2016), "Guide for recommended practices to perform crack tip opening displacement tests in high strength low alloy steels", *Soldagem and Inspeção*, 21, (3), 290–302.

Bain, E. C. and Paxton, H. W. (1966), *Alloying Elements in Steel*, Metals Park, OH: American Society for Metals.

Baji, H. Yang, W. and Li, C. Q. (2017), "An optimum strengthening strategy for corrosion affected reinforced concrete structures", *ACI Structural Journal*, 114, (6), 1591–1602.

Baji, H., Yang, W. and Li, C. Q. (2018), "Optimal FRP-strengthening strategy for corrosion-affected reinforced concrete columns", *Structure and Infrastructure Engineering*, 14, (12), 1586–1597.

Baji, H., Yang, W., Li, C. Q. and Shi, W. H., (2020), "A probabilistic model for time to cover cracking due to corrosion", *Structural Concrete*, 12. 201900341.

Bammou, L., Belkhaouda, M., Salghi, R., Benali, O., Zarrouk, A., Zarrok, H. and Hammouti, B. (2014), "Corrosion inhibition of steel in sulfuric acidic solution by the Chenopodium Ambrosioides Extracts", *Journal of the Association of Arab Universities for Basic and Applied Sciences*, 16, 83–90.

Bandara, C., Dissanayake, U. and Dissanayake, P. (2015), "Novel method for developing sn curves for corrosion fatigue damage assessment of steel structures", *6th International Conference on Structural Engineering and Construction Management*, Kandy.

Beech, I. B., Cheung, C. S., Chan, C. P., Hill, M. A., Franco, R. and Lino, A. R. (1994), "Study of parameters implicated in the biodeterioration of mild steel in the presence of different species of sulphate-reducing bacteria", *International Biodeterioration and Biodegradation*, 34, (3), 289–303.

Beech, I. B. and Sunner, J. (2004), "Biocorrosion: Towards understanding interactions between biofilms and metals", *Current Opinion in Biotechnology*, 15, (3), 181–186.

Beese, P., Venzlaff, H., Srinivasan, J., Garrelfs, J., Stratmann, M. and Mayrhofer, K. J. (2013), "Monitoring of anaerobic microbially influenced corrosion via electrochemical frequency modulation", *Electrochimica Acta*, 105, 239–247.

Beidokhti, B., Dolati, A. and Koukabi, A. (2009). "Effects of alloying elements and microstructure on the susceptibility of the welded HSLA steel to hydrogen-induced cracking and sulfide stress cracking", *Materials Science and Engineering: A*, 507, (1), 167–173.

Bell, R. G. and Lim, C. K. (1981), "Corrosion of mild and stainless steel by four tropical Desulfovibrio Desulfuricans strains", *Canadian Journal of Microbiology*, 27, (2), 242–245.

Belmokre, K., Azzouz, N., Kermiche, F., Wery, M. and Pagetti, J. (1998), "Corrosion study of carbon steel protected by a primer, by electrochemical impedance spectroscopy (EIS) in 3% NaCl medium and in a soil simulating solution", *Materials and Corrosion*, 49, (2), 108–113.

Belmonte, H. M. S., Mulheron, M. J. and Smith, P. A. (2009), "Some observations on the strength and fatigue properties of samples extracted from cast iron water mains", *Fatigue and Fracture of Engineering Materials and Structures*, 32, (11), 916–925.

Belonoshko, A. B. and Ahuja, R. (2017), "Embedded-atom molecular dynamic study of iron melting", *Physics of the Earth and Planetary Interiors*, 102, (3–4), 171–184.

Benjamin, A. C., Freire, J. L. F. and Vieira, R. D. (2007), "Part 6: Analysis of pipeline containing interacting corrosion defects", *Experimental Techniques*, 31, (3), 7482.

Benmoussa, A., Hadjel, M. and Traisnel, M. (2006), "Corrosion behavior of API 5L X-60 pipeline steel exposed to near-neutral pH soil simulating solution", *Materials and Corrosion*, 57, (10), 771–777.

Bentiss, F., Traisnel, M., Gengembre, L. and Lagrenée, M. (1999), "A new triazole derivative as inhibitor of the acid corrosion of mild steel: Electrochemical studies, weight loss determination, SEM and XPS", *Applied Surface Science*, 152, (3), 237–249.

Berardi, L., Giustolisi, O., Kapelan, Z. and Savic, D. (2008), "Development of pipe deterioration models for water distribution systems using EPR", *Journal of Hydroinformatics*, 10, (2), 113–126.

Bhadeshia, H. K. D. H. (2016), "Prevention of hydrogen embrittlement in steels", *ISIJ International*, 56, (1), 24–36.

Bhandari, J., Khan, F., Abbassi, R., Garaniya, V. and Ojeda, R. (2015), "Modelling of pitting corrosion in marine and offshore steel structures—A technical review", *Journal of Loss Prevention in the Process Industries*, 37, 39–62.

Booth, G. (1964), "Sulphur bacteria in relation to corrosion", *Journal of Applied Bacteriology*, 27, 174–181.

Bradley, W. L., and Srinivasan, M. N. (1990), "Fracture and fracture toughness of cast irons", *International materials reviews*, 35, (1), 129–116.

Bramfitt, B. L. (1998), "Structure/property relationships in irons and steels", *ASM International*, 153–173.

British Standards Institution. (1980), *BS 5400.10 Steel, Concrete and Composite Bridges, Part 10: Code of Practice for Fatigue*, London: British Standards Institution.

British Standards Institution. (1991), *BS 7448-1 Fracture Mechanics Toughness Tests. Part 1: Method for Determination of K_{IC}, Critical CTOD and Critical J Values of Metallic Materials*, British Standards Instituation.

British Standards Institution. (2014), *BS 7608 Guide to Fatigue Design and Assessment of Steel Products*, British Standards Institution.

Buehler, M. J. and Gao, H. (2006), "Dynamical fracture instabilities due to local hyperelasticity at crack tips", *Nature*, 439, (7074), 307.

Cabrini, M., Lorenzi, S., Marcassoli, P. and Pastore, T. (2011), "Hydrogen embrittlement behavior of HSLA line pipe steel under cathodic protection", *Corrosion Reviews*, 29, (5–6), 261–274.

Caleyo, F., González, J. L. and Hallen, J. M. (2002), "A study on the reliability assessment methodology for pipelines with active corrosion defects", *International Journal of Pressure Vessels and Piping*, 79, (1), 77–86.

Caleyo, F., Velázquez, J. C., Valor, A. and Hallen, J. M. (2009), "Probability distribution of pitting corrosion depth and rate in underground pipelines: A Monte Carlo study", *Corrosion Science*, 51, (9), 1925–1934.

Calleja, G., Botas, J. A., Sánchez-Sánchez, M. and Orcajo, M. G. (2010), "Hydrogen adsorption over Zeolite-like MOF materials modified by ion exchange", *International Journal of Hydrogen Energy*, 35, (18), 9916–9923.

Campbell, J. (2003), *Castings*, London: Butterworth-Heinemann.

Cao, J. and Lin, J. (2008), "A study on formulation of objective functions for determining material models", *International Journal of Mechanical Sciences*, 50, (2), 193–204.
Caproco Corrosion Prevention Ltd. (1985), *Underground Corrosion of Water Pipes in Canadian Cities Case: The City of Calgary*, Report prepared for CANMET, Ottawa.
Casanova, T., Soto, F., Eyraud, M. and Crousier, J. (1997), "Hydrogen absorption during zinc plating on steel", *Corrosion Science*, 39, 529–537.
Cerit, M., Genel, K. and Eksi, S. (2009), "Numerical investigation on stress concentration of corrosion pit", *Engineering Failure Analysis*, 16, (7), 2467–2472.
Chalaftris, G. (2003), *Evaluation of Aluminium-Based Coatings for Cadmium Replacement* (Doctoral thesis), Bedford: Cranfield University.
Chen, F. J, Li, C. Q., Baji, H. and Ma, B. G. (2020), "Quantification of non-uniform distribution and growth of corrosion products at steel-concrete interface", *Construction and Building Materials*, 237, 117610.
Chen, W. F. & Duan, L. (2014). *Bridge engineering handbook: Construction and maintenance.* CRC press.
Cheng, Y. (2007a), "Analysis of electrochemical hydrogen permeation through X-65 pipeline steel and its implications on pipeline stress corrosion cracking", *International Journal of Hydrogen Energy*, 32, 1269–1276.
Cheng, Y. (2007b), "Fundamentals of hydrogen evolution reaction and its implications on near-neutral Ph stress corrosion cracking of pipelines", *Electrochimica Acta*, 52, (7), 2661–2667.
Chilingar, G. V., Mourhatch, R. and Al-Qahtani, G. D. (2013), *The Fundamentals of Corrosion and Scaling for Petroleum and Environmental Engineers*, Houston: Gulf Publishing Company.
Chisholm, C., Mook, W., Bufford, D. C., Hattar, K. M., Jungjohann, K. L., Hayden, ... Ostraat, M. (2016), *Steel Corrosion Mechanisms during Pipeline Operation: In-situ Characterization* (No. SAND2016-8199C), Office of Scientific and Technical Information, US Department of Energy.
Cole, I. S. and Marney, D. (2012), "The science of pipe corrosion: A review of the literature on the corrosion of ferrous metals in soils", *Corrosion Science*, 56, 5–16.
Cownie, A. and Palmer, L. S. (1952), "The effect of moisture on the electrical properties of soil", *Proceedings of the Physical Society. Section B*, 65, (4), 295.
Cramer, S. D. and Covino, B. S. (Eds.). (2003), *Corrosion: Fundamentals, Testing, and Protection* (Vol. 13A), Materials Park, OH: ASM International.
Cramer, S. D., Covino, B. S. and Moosbrugger, C. (2005), *ASM Handbook Volume 13b: Corrosion: Materials* (Vol. 13), Materials Park, OH: ASM International.
Cronin, D. S. (2000), *Assessment of Corrosion Defects in Pipelines* (Doctoral thesis), Waterloo: University of Waterloo.
Cronin, D. S. and Pick, R. J. (2002), "Prediction of the failure pressure for complex corrosion defects", *International Journal of Pressure Vessels and Piping*, 79, (4), 279–287.
Dafter, M. R. (2014), *Electrochemical Testing of Soils for Long-Term Prediction of Corrosion of Ferrous Pipes*, Doctor of Philosophy, Callaghan: The University of Newcastle.
Danford, M. D. (1983), *An Electrochemical Method or Determining Hydrogen Concentrations in Metals and Some Applications*, NASA technical paper 2113.
Darmawan, M. S., Refani, A. N., Irmawan, M., Bayuaji, R. and Anugraha, R. B. (2013), "Time dependent reliability analysis of steel I bridge girder designed based on SNI T-02-2005 and SNI T-3-2005 subjected to corrosion", *Procedia Engineering*, 54, 270–285.
Davalos, J., Gracia, M., Marco, J. F. and Gancedo, J. R. (1992), "Corrosion of weathering steel and iron under wet-dry cycling conditions: Influence of the rise of temperature during the dry period", *Hyperfine Interactions*, 69, 871–874.
Davis, J. R. (Ed.). (2000), *Corrosion: Understanding the Basics.* Material Parks, OH: ASM International.

De la Fuente, D., Alcántara, J., Chico, B., Díaz, I., Jiménez, J. A. and Morcillo, M. (2016), "Characterisation of rust surfaces formed on mild steel exposed to marine atmospheres using XRD and SEM/Micro-Raman techniques", *Corrosion Science*, 110, 253–264.

De la Fuente, D., Díaz, I., Simancas, J., Chico, B. and Morcillo, M. (2011), "Long-term atmospheric corrosion of mild steel", *Corrosion Science*, 53, (2), 604–617.

Deb, P. & Chaturvedi, M. (1985), "Thermomechanical treatment of AISI 1015 Steel", *Materials Science and Engineering*, 68, (2), 207–217.

Delaunois, F., Tshimombo, A., Stanciu, V. and Vitry, V. (2016), "Monitoring of chloride stress corrosion cracking of austenitic stainless steel: Identification of the phases of the corrosion process and use of a modified accelerated test", *Corrosion Science*, 110, 273–283.

Denison, I. and Hobbs, R. (1934), "Corrosion of ferrous metals in acid soils", *Journal of Research, National Bureau of Standards*, 13, 125.

Denison, I. A. and Romanoff, M. (1954), "Corrosion of nickel cast irons in soils", *Corrosion*, 10, (6), 199–204.

Devore, J. L. (2012), *Probability and Statistics for Engineering and the Sciences* (8th ed.), Boston, MA: Brooks/Cole and Cengage Learning.

Di Tommaso, D., Hernandez, S. E. R., Du, Z. and de Leeuw, N. H. (2012), "Density functional theory and interatomic potential study of structural, mechanical and surface properties of calcium oxalate materials", *RSC Advances*, 2, (11), 4664–4674.

Digges, T. G., Rosenberg, S. J. and Geil, G. W. (1966), *Heat Treatment and Properties of Iron And Steel (No. NBS-MONO-88)*, Gaithersburg, MD: National Bureau of Standards.

Djukic M. B., Bakic G. M., Zeravcic V. S., Sedmak A. and Rajicic B. (2016), "Hydrogen embrittlement of industrial components: prediction, prevention, and models", *Corrosion*, 72, (7), 943–961.

Doran, D. and Cather, B. (2013), *Construction Materials Reference Book*, London: Routledge.

Doyle, G., Seica, M. V. and Grabinsky, M. W. (2003), "The role of soil in the external corrosion of cast iron water mains in Toronto", Canada. *Canadian Geotechnical Journal*, 40, (2), 225–236.

Dražić, D. M. and Vašĉiž, V. (1989), "The correlation between accelerated laboratory corrosion tests and atmospheric corrosion station tests on steels", *Corrosion Science*, 29, (10), 1197–1204.

Eggum, T. J. (2013), *Hydrogen in Low Carbon Steel: Diffusion, Effect on Tensile Properties, and an Examination of Hydrogen's Role in the Initiation of Stress Corrosion Cracking in a Failed Pipeline* (Doctoral thesis), Calgary: University of Calgary.

Enning, D. (2012), *Bio-electrical Corrosion of Iron by Lithotrophic Sulfate-Reducing Bacteria*, Ph.D. thesis, Bremen: University of Bremen.

Eslami, A., Fang, B., Kania, R., Worthingham, B., Been, J., Eadie, R. and Chen, W. (2010), "Stress corrosion cracking initiation under the disbonded coating of pipeline steel in near-neutral Ph environment", *Corrosion Science*, 52, 3750–3756.

Eslami, A., Kania, R., Worthingham, B., Boven, G. V., Eadie, R. and Chen, W. (2011), "Effect of CO_2 and R-ratio on near-neutral Ph stress corrosion cracking initiation under a disbonded coating of pipeline steel", *Corrosion Science*, 53, 2318–2327.

Evans, U. R. and Taylor, C. A. J. (1972), "Mechanism of atmospheric rusting", *Corrosion Science*, 12, 227–246.

Fahimi, A., Evans, T. S., Farrow, J., Jesson, D. A., Mulheron, M. J. and Smith, P. A. (2016), "On the residual strength of aging cast iron trunk mains: Physically-based models for asset failure", *Materials Science and Engineering*: *A*, 663, 204–212.

Fang, B. Y., Han, E. H., Wang, J. Q. and Ke, W. (2007), "Stress corrosion cracking of X-70 pipeline steel in near neutral pH solution subjected to constant load and cyclic load testing", *Corrosion Engineering, Science and Technology*, 42, (2), 123–129.

Fei, X., Li, M., Xu, H., Li, Y.-Q. and Cai, D. (2007), "Influence of soil humidity on corrosion behavior of X70 steel in yellow pebble soil", *Corrosion Science and Protection Technology*, 19, 35.

Ferguson, P. h. and Nicholas, D. M. F. (1991), "External corrosion of buried iron and steel water mains", *ACA Annual Conference*, 17, (4), 7–10.

Ferhat, M., Benchettara, A., Amara, S. and Najjar, D. (2014), "Corrosion behaviour of Fe-C alloys in a sulfuric medium", *Journal of Materials and Environmental Science*, 5, (4), 1059–1068.

Finšgar, M. and Jackson, J. (2014), "Application of corrosion inhibitors for steels in acidic media for the oil and gas industry: A review", *Corrosion Science*, 86, 17–41.

Firouzi, A., Yang, W., Li, C. Q. and Yang, S. T. (2020), "Failure of corrosion affected buried cast iron pipes subject to water hammer", *Engineering Failure Analysis*, 118, 104993.

Fitzgerald, J. H. (1968), "Corrosion as a primary cause of cast-iron main breaks", *Journal American Water Works Association*, 60, (8), 882–897.

Flitton, M. K. A. and Escalante, E. (2003), Simulated service testing, In S. D. Cramer and B. S. Covino (eds.), *Corrosion: Fundamentals, Testing, and Protection* (pp. 497–500), *ASM Handbook* (Vol. 13A), Materials Park, OH: ASM International.

Folkman, S. (2018), *Water Main Break Rates in the USA and Canada: A Comprehensive Study* (Paper No. 174), Logan, UT: Utah State University.

Foroulis, Z. A., & Uhlig, H. H. (1965), "Effect of impurities in iron on corrosion in acids", *Journal of The Electrochemical Society*, 112(12), 1177.

Frankel, G. S. (2003), Pitting corrosion. In S. D. Cramer and B. S. Covino (eds.), *Corrosion: Fundamentals, Testing, and Protection* (pp. 236–241), *ASM Handbook* (Vol. 13A), Materials Park, OH: ASM International.

Freire, J. L. F., Vieira, R. D., Castro, J. T. P. and Benjamin, A. C. (2007), "Part 5: Rupture tests of pipeline containing complex-shaped metal loss defects", *Experimental Techniques*, 31, (2), 57–62.

Fu, G. Y., Yang, W., Deng, W. N., Li, C. Q. and Setunge, S. (2019), "Prediction of fracture failure of steel pipes with sharp corrosion pits using time-dependent reliability method with lognormal process", *ASME Journal of Pressure Vessel Technology*, 141, 031401-1.

Gao, H., Li, C. Q., Wang, W. G., Wang, Y. L. and Zhang, B. H. (2020), "Factors affecting the agreement between unloading compliance method and normalization method", *Engineering Fracture Mechanics*, 235, 107146.

Garbatov, Y., Soares, C. G., Parunov, J. and Kodvanj, J. (2014), "Tensile strength assessment of corroded small scale specimens", *Corrosion Science*, 85, 296–303.

Garet, M., Brass, A., Haut, C. and Guttierez-Solana, F. (1998), "Hydrogen trapping on non-metallic inclusions in Cr-Mo low alloy steels", *Corrosion Science*, 40, (7), 1073–1086.

Gaylarde, C. C. (1992), "Sulphate-reducing bacteria which do not induce accelerated corrosion", *International Biodeterioration and Biodegradation*, 30, (4), 331–338.

Gerhardus, H. K. (2001), "Tests for stress-corrosion cracking", *Advanced Materials and Processes*, 159, (8), 36–38.

Gerhold, W. (1976), "Corrosion behavior of ductile cast-iron pipe in soil environments", *AWWA Journal*, 68, (12), 674–678.

GHD. (2015), *Infrastructure Maintenance, A Report for Infrastructure Australia*, Canberra: Infrastructure Australia.

Gonzaga, R. A. (2013), "Influence of ferrite and pearlite content on mechanical properties of ductile cast irons", *Materials Science and Engineering: A*, 567, 1–8.

Goodman, N., Muster, T., Davis, P., Gould, S. and Marney, D. (2013, November). Accelerated test based on EIS to predict buried steel pipe corrosion. *Paper Presented at the 2013 Corrosion and Prevention Conference*, Brisbane, Australia.

Graedel, T. E. and Frankenthal, R. P. (1990), "Corrosion mechanisms for iron and low alloy steels exposed to the atmosphere", *Journal of the Electrochemical Society*, 137, (8), 2385–2394.

Gupta, S. K. and Gupta, B. K. (1979), "The critical soil moisture content in the underground corrosion of mild steel", *Corrosion Science*, 19, (3), 171–178.

Gutman E. M. (1989), *Mechanochemistry and Corrosion Protection of Metal*, Beijing: Science Press.

Hagiwara, N. and Oguchi, N. (1999), "Fracture toughness of line pipe steels under cathodic protection using crack tip opening displacement tests", *Corrosion*, 55, (5):503–511.

Hamilton, W. A. (1985), "Sulphate-reducing bacteria and anaerobic corrosion", *Annual Reviews in Microbiology*, 39, 195–217.

Hardie, D., Charles, E. A. and Lopez, A. H. (2006), "Hydrogen embrittlement of high strength pipeline steels", *Corrosion Science*, 48, (12), 4378–4385.

Hassel A. W, Stratmann, M. and Widdel, F. (2012), "Marine sulfate-reducing bacteria cause serious corrosion of iron under electroconductive biogenic mineral crust", *Environmental Microbiology*, 14, (7), 1772–1787.

Haynes R. (2010) *Sydney Harbour Bridge rusting despite relentless paint job*, DailyTelegraph. Reviewed on Nov 2019.

Head, K. H. and Epps, R. (1986), *Manual of Soil Laboratory Testing* (Vol. 3, pp. 798–869), London: Pentech Press.

Hejazi, D., Calka, A., Dunne, D. and Pereloma, E. (2016), Effect of gaseous hydrogen charging on the tensile properties of standard and medium Mn X70 pipeline steels. *Materials Science and Technology*, 32, (7), 675–683.

Hertzberg, R. W. (1996), *Deformation and Fracture Mechanics of Engineering Materials* (4th ed.), New York, NY: Wiley and Sons.

Higuchi, M. and Iida, K. (1991), "Fatigue strength correction factors for carbon and low-alloy steels in oxygen-containing high-temperature water", *Nuclear Engineering and Design*, 129, 293–306.

Horner, D. A., Connolly, B. J., Zhou, S., Crocker, L. and Turnbull, A. (2011), "Novel images of the evolution of stress corrosion cracks from corrosion pits", *Corrosion Science*, 53, (11), 3466–3485.

Hou, Y., Lei, D., Li, S., Yang, W. and Li, C. Q. (2016), "Experimental investigation on corrosion effect on mechanical properties of buried metal pipes", *International Journal of Corrosion*, 2016, (6), 1–13.

Hu, C., Xia, S., Li, H., Liu, T., Zhou, B., Chen, W. and Wang, N. (2011), "Improving the intergranular corrosion resistance of 304 stainless steel by grain boundary network control", *Corrosion Science*, 53, 1880–1886.

Hu, J., Shun-An, C. and Xie, J. (2013), "EIS study on the corrosion behavior of rusted carbon steel in 3% NaCl solution", *Anti - Corrosion Methods and Materials*, 60, 100–105.

Huang, H. and Shaw, W. J. D. (1995), "Hydrogen embrittlement interactions in cold-worked steel", *Corrosion*, 51, (1), 30–36.

Hubert, C. (2009), "Turbulent air-water flows in hydraulic structures: dynamic similarity and scale effects", *Environ Fluid Mech*, 9,(2), 125–142.

Iacoviello, F., Galland, J. and Habashi, M. (1998), "A thermal outgassing method (T.O.M.) to measure the hydrogen diffusion coefficients in austenitic, austeno-ferritic and ferritic-perlitic steels", *Corrosion Science*, 40, (8), 1281–1293.

Ichitani, K. and Kanno, M. (2003), "Visualization of hydrogen diffusion in steels by high sensitivity hydrogen microprint technique", *Science and Technology of Advanced Material*, 4, 545–551.

International Organization for Standardization. (2000), *BS 410-1:2000 Test sieves: Technical Requirements and Testing. Test Sieves of Metal Wire Cloth* (4th ed.), London: British Standards Institution.

International Organization for Standardization. (2012), *ISO 9224 Corrosion of Metals and Alloys. Corrosivity of Atmospheres. Guiding Values for the Corrosivity Categories*, Geneva, International Organization for Standardization.

Jack, T. R. (2002), "Biological corrosion failures, ASM handbook", *Failure Analysis and Prevention*, 882–890.

Jakobs, J. A. and Hewes, F. W. (1987), "Underground corrosion of water pipes in Calgary, Canada", *Materials Performance*, 36, (5), 42–49.

Jakubowski, M. (2011), "Influence of pitting corrosion on strength of steel ships and offshore structures", *Polish Maritime Research*, 18, (4), 54–58.

Javaherdashti, R. (1999), "A review of some characteristics of MIC caused by sulfate-reducing bacteria: past, present and future", *Anti-corrosion Methods and Materials*, 46, 173–180.

Javaherdashti, R., Raman, R. S., Panter, C. and Pereloma, E. (2006), "Microbiologically assisted stress corrosion cracking of carbon steel in mixed and pure cultures of sulfate reducing bacteria", *International Biodeterioration and Biodegradation*, 58, 27–35.

Javed, M. A., Stoddart, P. R., Palombo, E. A., Mcarthur, S. L. and Wade, S. A. (2014), "Inhibition or acceleration: Bacterial test media can determine the course of microbiologically influenced corrosion", *Corrosion Science*, 86, 149–158.

Javed, M. A., Stoddart, P. R. and Wade, S. A. (2015), "Corrosion of carbon steel by sulphate reducing bacteria: Initial attachment and the role of ferrous ions", *Corrosion Science*, 93, 48–57.

Ji, J., Zhang, C., Kodikara, J. and Yang, S. Q. (2015), "Prediction of stress concentration factor of corrosion pits on buried pipes by least squares support vector machine", *Engineering Failure Analysis*, 55, 131–138.Junior, A. M. J., Guedes, L. H. and Balancin, O. (2012), "Ultra grain refinement during the simulated thermomechanical-processing of low carbon steel", *Journal of Materials Research and Technology*, 1, (3), 141–147.

Kaesche, H. (2011), *Corrosion of Metals: Physicochemical Principles and Current Problems*, London: Springer.

Kayser, J. R. and Nowak, A. S. (1989). "Capacity loss due to corrosion in steel-girder bridges". *Journal of Structural Engineering*, 115, (6), 1525–1537.

Keil, B. and Devletian, J. (2011), Comparison of the mechanical properties of steel and ductile iron pipe materials, In D. H. S Jeong (ed.), *Pipelines 2011: A Sound Conduit for Sharing Solutions* (pp. 1301–1312), Reston, VA: American Society of Civil Engineers.

Kentish, P. (2007), "Stress corrosion cracking of gas pipelines - Effect of surface roughness, orientations and flattening", *Corrosion Science*, 49, 2521–2533.

Kim, J., Bae, C., Woo, H., Kim, J. and Hong, S. (2007), Assessment of residual tensile strength on cast iron pipes, In L. Osborn and M. Najafi (eds.), *Pipelines 2007: Advances and Experiences with Trenchless Pipeline Projects* (pp. 1–7), Reston, VA: American Society of Civil Engineers.

Kirby, P. (1977), *Internal Corrosion and Loss of Strength of Iron Pipes*, Swindon: Water Research Centre.

Kolios, A., Srikanth, S. and Salonitis, K. (2014), "Numerical simulation of material strength deterioration due to pitting corrosion", *Procedia CIRP*, 13, 230–236.

Kreysa, G. and Schütze, M. (Eds.) (2008), *Corrosion Handbook* (2nd ed.), London: John Wiley & Sons Inc.

Kritzer, P. (2004), "Corrosion in high-temperature and supercritical water and aqueous solutions, a review", *The Journal of Supercritical Fluids*, 29, 1–29.

Kucera, V. and Mattsson, E. (1987), "Atmospheric corrosion", *Corrosion Mechanisms*, 28, 211–284.

Kuch, A. (1988), "Investigations of the reduction and re-oxidation kinetics of iron (III) oxide scales formed in waters", *Corrosion Science*, 28, 221–231.

Landolfo, R., Cascini, L. and Portioli, F. (2010), "Modeling of metal structure corrosion damage: A state of the art report", *Sustainability*, 2, (7), 2163–2175.

Laque, F. L. (1958), "The corrosion resistance of ductile iron", *Corrosion*, 14, (10), 55–62.

Lau, I. (2018), *Steel–Concrete Interface and Its Influence on Concrete Cracking in Reinforced Concrete* (Doctoral thesis), Melbourne: RMIT University.

Lau, I., Fu, G. Y., Li, C. Q., De Silva, S and Guo, Y. F. (2018), "Critical crack depth in corrosion-induced concrete cracking", *ACI Structures Journal*, 115, (4), 1175–1184.

Lau, I., Li, C. Q. and Chen, F. J. (2020), "Analytical and experimental investigation on corrosion-induced concrete cracking", *International Journal of Civil Engineering*, 18, (1), 99–112.

Lau, I., Li, C. Q. and Fu, G. Y. (2019), "Prediction of time to corrosion-induced concrete cracking based on fracture mechanics criteria", *ASCE Journal of Structural Engineering*, 145, (8), 04019069.

Lee W., Lewandowski Z., Morrison M., Characklis W. G., Avci R. and Nielsen P. H. (1993), "Corrosion of mild steel underneath aerobic biofilms containing sulfate-reducing bacteria. Part I: at high dissolved oxygen concentrations", *Biofouling*, 7, 217–239.

Lee, W., Lewandowski, Z., Nielsen, P. H. and Hamilton, W. A. (1995), "Role of sulfate reducing bacteria in corrosion of mild steel: A review", *Biofouling*, 8, (3), 165–194.

Lehockey, E. M., Brennenstuhl, A. M. and Thompson, I. (2004), "On the relationship between grain boundary connectivity, coincident site lattice boundaries, and intergranular stress corrosion cracking", *Corrosion Science*, 46, (10), 2383–2404.

Li, C. Q. (1999), "Modelling time-to-corrosion cracking in chloride contaminated reinforced concrete structures – discussion", *ACI Materials Journal*, 96, (5), 611–613.

Li, C. Q. (2000), "Corrosion initiation of reinforcing steel in concrete under natural salt spray and service loading – results and analysis", *ACI Materials Journal*, 97, (6), 690–697.

Li, C. Q. (2001), "Initiation of chloride induced reinforcement corrosion in concrete structural members – experimentation", *ACI Structural Journal*, 98, (4), 501–510.

Li, C. Q. (2002), "Initiation of chloride induced reinforcement corrosion in concrete structural members – prediction", *ACI Structural Journal*, 99, (2), 133–141.

Li, C. Q. (2003a), "Life cycle modelling of corrosion affected concrete structures – propagation", *ASCE Journal of Structural Engineering*, 129, (6), 753–761.

Li, C. Q. (2003b), "Life cycle modelling of corrosion affected concrete structures – initiation", *ASCE Journal of Materials in Civil Engineering*, 15, (6), 594–601.

Li, C. Q. (2004), "Reliability based service life prediction of corrosion affected concrete structures", *ASCE Journal of Structural Engineering*, 130, (10), 1570–1577.

Li, C. Q. (2005), "Time dependent reliability analysis of the serviceability of corrosion affected concrete structures", *Journal of Material and Structural Reliability*, 3, (2), 105–116.

Li, C. Q. (2018), "Corrosion and Its Effects on Deterioration and Remaining Safe Life of Civil Infrastructure", *6th International Symposium on Life Cycle Civil Engineering (IALCEE2018), Keynote Paper*, 28–31 October, 2018, Ghent.

Li, L. (2018), Service Life Prediction of Bridges Subjected to Corrosion Using *Time-Dependent Reliability Method* (Doctoral thesis), Melbourne: RMIT University.

Li, C. Q., Baji, H. and Yang, W. (2019a), "Optimal inspection plan for deteriorating structures using stochastic models", *ASCE Journal of Structural Engineering*, 145, (11), 04019119.

Li, M. and Cheng, Y. (2008), "Corrosion of the stressed pipe steel in carbonate–bicarbonate solution studied by scanning localized electrochemical impedance spectroscopy", *Electrochimica Acta*, 53, 2831–2836.

Li, C. Q., Firouzi, A. and Yang, W. (2016a), "Closed-form solution to first passage probability for nonstationary lognormal processes", *ASCE Journal of Engineering Mechanics*, 142, (12), 04016103.

Li, C. Q., Firouzi, A. and Yang, W. (2017), "Prediction of pitting corrosion-induced perforation of ductile iron pipes", *ASCE Journal of Engineering Mechanics*, 143, (8), 04017048.

Li, C. Q., Fu, G. and Yang, W. (2016b), "Stress intensity factors for inclined external surface cracks in pressurized pipes", *Engineering Fracture Mechanics*, 165, 72–86.

Li, S., Jung, S., Park, K. W., Lee, S.-M. and Kim, Y. G. (2007a), "Kinetic study on corrosion of steel in soil environments using electrical resistance sensor technique", *Materials Chemistry and Physics*, 103, 9–13.

Li, S. Y., Kim, Y. G., Jeon, K. S., Kho, Y. T. and Kang, T. (2001), "Microbiologically influenced corrosion of carbon steel exposed to anaerobic soil", *Corrosion*, 57, 815–828.

Li, C. Q., Lawanwisut, W. and Zheng, J. J. (2004), "Service life modelling of corrosion affected concrete structures", *Journal of Material and Structural Reliability*, 2, (2), 105–121.

Li, C. Q., Lawanwisut, W. and Zheng, J. J. (2005a), "A time dependent reliability method to assess the serviceability of corrosion affected concrete structures", *ASCE Journal of Structural Engineering*, 131, (11), 1674–1680.

Li, L, Li, C. Q. and Mahmoodian, M. (2019b), "Effect of applied stress on corrosion and mechanical property of mild steel", *ASCE Journal of Materials in Civil Engineering*, 31, (2), 04018375.

Li, L, Li, C. Q. and Mahmoodian, M. (2019c), "Prediction of fatigue failure of steel beams subjected to simultaneous corrosion and cyclic loading", *Structures*, 19, 386–393.

Li, L, Li, C. Q., Mahmoodian, M. and Shi, W. H. (2020a), "Corrosion induced degradation of fatigue strength of steel in-service for 128 years", *Structures*, 23, 415–424.

Li, M., Lin, H. and Cao, C. (2000), "Influence of moisture content on soil corrosion behavior of carbon steel", *Corrosion Science and Protection Technology*, 12, 218–220.

Li, C. Q., Mackie, R. I. and Lawanwisut, W. (2007b), "A risk cost optimized maintenance strategy for corrosion affected concrete structures", *Journal of Computer-Aided Civil and Infrastructure Engineering*, 22, 335–346.

Li, C. Q. and Mahmoodian, M. (2013), "Risk based service life prediction of underground cast iron pipes subjected to corrosion", *Reliability Engineering and System Safety*, 119, 102–108.

Li, L., Mahmoodian, M. and Li, C. Q. (2020b), "Prediction of fatigue life of corrosion affected riveted connections in steel structures", *Structure and Infrastructure Engineering*, 16, (11), 1524–1538.

Li, L, Mahmoodian, M, Li, C. Q. and Robert, D. (2018a), "Effect of corrosion and hydrogen embrittlement on microstructure and mechanical properties of mild steel", *Construction and Building Materials*, 170, 78–90.

Li, L, Mahmoodian, M, Wasim, M. and Li, C. Q. (2018b), "Preferred corrosion and its effect on steel delamination", *Construction and Building Materials*, 193, 576–588.

Li, C. Q. and Melchers, R. E. (2005a), "Time dependent risk assessment of structural deterioration caused by reinforcement corrosion", *ACI Structural Journal*, 102, (5), 754–762.

Li, C. Q. and Melchers, R. E. (2005b), "Time-dependent reliability analysis of corrosion-induced concrete cracking", *ACI Structural Journal*, 102, (4), 543–549.

Li, C. Q. and Melchers, R. E. (2006), "Time dependent serviceability of corrosion affected concrete structures", *Magazine of Concrete Research*, 58, (9), 567–574.

Li, C. Q., Melchers, R. E. and Lawanwisut, W. (2005b), "Vulnerability assessment of corrosion affected concrete structures", *Magazine of Concrete Research*, 57, (9), 557–565.

Li, C. Q., Melchers, R. E. and Zheng, J. J. (2006), "Analytical model for corrosion induced crack width in reinforced concrete structures", *ACI Structural Journal*, 103, (4), 479–487.

Li, Q. X., Wang, Z. Y., Han, W. and Han, E. H. (2008a), "Characterization of the rust formed on weathering steel exposed to Qinghai salt lake atmosphere", *Corrosion Science*, 50, 365–371.

Li, C. Q., Wang, F., Li, L. and Li, Y. R. (2020c), "Investigating effect of corrosion on mechanical properties based on macro and atomic scale studies", *Proc. of ICNASA, Keynote Paper*, 31 January – 3 February, 2020, Melbourne.

Li, C. Q., Wang, F., Li, L. and Li, Y.R. (2020d), "Investigating effect of corrosion on mechanical properties based on macro & atomic scale studies", Keynote paper at ICNASA on Jan 31-Feb 3, 2020 in Melbourne.

Li, C. Q. and Yang, S. T. (2011), "Prediction of concrete crack width under combined reinforcement corrosion and applied load", *ASCE Journal of Engineering Mechanics*, 137, (11), 722–731.

Li, C. Q. and Yang, S. T. (2012), "Stress intensity factors for high aspect ratio semi-elliptical internal surface cracks in pipes", *International Journal of Pressure Vessels and Piping*, 96–97, 13–23.

Li, C. Q., Yang, Y and Melchers, R. E. (2008b), "Prediction of reinforcement corrosion in concrete and its effects on concrete cracking and strength reduction", *ACI Materials Journal*, 105, (1), 3–10.

Li, C. Q., Yang, S. T. and Saafi, M. (2014), "Numerical simulation of behaviour of reinforced concrete structures considering corrosion effects on bonding", *ASCE Journal of Structural Engineering*, 140, (12), 04014092.

Li, C. Q. Yang, W. and Shi, W. H. (2019d), "Corrosion effect of ferrous metals on degradation and remaining service life of infrastructure using pipe fracture as example", *Structure and Infrastructure Engineering*, 15, 1–16.

Li, S. X., Yu, S. R., Zeng, H. L., Li, J. H. and Liang, R. (2009), "Predicting corrosion remaining life of underground pipelines with a mechanically-based probabilistic model", *Journal of Petroleum Science and Engineering*, 65, (3–4), 162–166.

Li, F., Yuan, Y. and Li, C. Q. (2011), "Corrosion propagation of prestressing steel strands in concrete subject to chloride attack", *Construction and Building Materials*, 25, 3878–3885.

Li, C. Q. and Zheng, J. J. (2005), "Propagation of reinforcement corrosion in concrete and its effects on structural deterioration", *Magazine of Concrete Research*, 57, (5), 262–271.

Li, C. Q., Zheng, J. J., Lawamwisut, W. and Melchers, R. E. (2007c), "Concrete delamination caused by steel reinforcement corrosion", *ASCE Journal of Materials in Civil Engineering*, 19, (7), 591–600.

Li, C. Q., Zheng, J. J. and Shao, L. (2003), "New solution for prediction of chloride ingress in reinforced concrete flexural members", *ACI Materials Journal*, 100, (4), 319–325. (Nominated for one of ACI Awards).

Li, J. B. and Zuo, J. E. (2008), "Influences of temperature and pH value on the corrosion behaviors of x80 pipeline steel in carbonate/bicarbonate buffer solution", *Chinese Journal of Chemistry*, 26, 1799–1805.

Lichtenstein, A. G. (1993), "The silver bridge collapse recounted. *Journal of Performance of Constructed Facilities*", 7, (4), 249–261.

Lino, R. E., Marins, Â. M. F., Marchi, L. A., Mendes, J. A., Penna, L. V., Neto, J. G. C. and Caldeira, J. H. P. (2017), "Influence of the chemical composition on steel casting performance", *Journal of Materials Research and Technology*, 6, (1), 50–56.

Little, B. J. and Lee, J. C. (2007), *Microbiologically Influenced Corrosion*, London: Wiley.

Liu, Z. Y., Li, X. G., Du, C. W. and Cheng, Y. F. (2009), "Local additional potential model for effect of strain rate on SCC of pipeline steel in an acidic soil solution", *Corrosion Science*, 51, 2863–2871.

Liu, Y., Wang, Z. and Ke, W. (2014a), "Study on influence of native oxide and corrosion products on atmospheric corrosion of pure Al", *Corrosion Science*, 80, 169–176.

Liu, T. M., Wu, Y. H., Luo, S. X. and Sun, C. (2010), "Effect of soil compositions on the electrochemical corrosion behavior of carbon steel in the simulated soil solution", *Materialwissenschaft und Werkstofftechnik*, 41, (4), 228–233.

Liu, Y., Xu, L. N., Zhu, J. Y. and Meng, Y. (2014b), "Pitting corrosion of 13Cr steel in aerated brine completion fluids", *Materials and Corrosion*, 65, (11), 1096–1102.

López, E., Osella, A. and Martino, L. (2006), "Controlled experiments to study corrosion effects due to external varying fields in embedded pipelines", *Corrosion Science*, 48, (2), 389–403.

Lu, Z. H., Ou, Y. B., Zhao, Y. G. and Li, C. Q. (2016), "Investigation of Corrosion of Steel Stirrups in Reinforced Concrete Structures", *Construction and Building Materials*, 127, 293–305.

Luo, H., Dong, C. F., Liu, Z. Y., Maha, M. T. J. and Li, X. G. (2013), "Characterization of hydrogen charging of 2205 duplex stainless steel and its correlation with hydrogen-induced cracking", *Materials and Corrosion*, 64, (1), 26–33.

Ma, F. Y. (2012), *Corrosive Effects of Chlorides on Metals*, Tech Open Access Publisher. https://www.intechopen.com/

Maalekian, M. (2007), "The effects of alloying elements on steels (I)", *Christian Doppler Laboratory for Early Stages of Precipitation*, 23, 221–230.

Mabuchi, K., Horn, Y., Takahashi, H. and Nagayama, M. (1991), "Effect of temperature and dissolved oxygen on the corrosion behavior of carbon steel in high-temperature water", *Corrosion*, 47, 500–508.

MacDonald, D. D. (1999), "Passivity–the key to our metals-based civilization", *Pure and Applied Chemistry,* 71, (6), 951–978.

Mahmoodian, M. and Li, C. Q. (2016), "Structural integrity of corrosion affected cast iron water pipes using a reliability based stochastic analysis method", *Structure and Infrastructure Engineering,* 12, (10), 1356–1363.

Mahmoodian, M. and Li, C. Q. (2017), "Failure assessment and safe life prediction of corroded oil and gas pipelines", *Journal of Petroleum Science and Engineering,* 151, 434–438.

Mahmoodian, M. and Li, C. Q. (2018), "Reliability based service life prediction of corrosion affected cast iron pipes considering multifailure modes", *ASCE Journal of Infrastructure Systems*, 24, (2), 04018004.

Man, O., Pantělejev, L. and Pešina, Z. (2009), "EBSD analysis of phase compositions of trip steel on various strain levels", *Materials Engineering*, 16, (2), 15–21.

Mangat, P. S. & Molloy, B. T. (1992), "Factors influencing chloride-induced corrosion of reinforcement in concrete", *Materials and Structures*, 25, (7), 404–411.

Mao, S. X. and Li M. (1998), "Mechanics and thermodynamics on the stress and hydrogen interaction in crack tip stress corrosion: Experiment and theory", *Journal of the Mechanics and Physics of Solids*, 46, (6), 1125–1137.

Marcus, P. (Ed.). (2011), *Corrosion Mechanisms in Theory and Practice* (3rd ed.), Boca Raton, FL: CRC Press.

Marcus, P. and Maurice, V. (2017), "Atomic level characterization in corrosion studies", *Philosophical Transactions of the Royal Society A: Mathematical, Physical and Engineering Sciences*, 375, (2017), 20160414.

Martin, C. (2013), "Corrosion of carbon steel under sequential aerobic–anaerobic environmental conditions", *Corrosion Science*, 76, 432–440.

MathWave Technologies. (2016). EasyFit Distribution Fitting Software (Version 5.6) [Computer software].

McDougall, J. (1966), "Microbial corrosion of metals", *Anti-Corrosion Methods and Materials*, 13, (8), 9–13.

Mcneill, L. S. and Edwards, M. (2002), "The importance of temperature in assessing iron pipe corrosion in water distribution systems", *Environmental Monitoring and Assessment*, 77, 229–242.

Melchers, R. E. (2013), "Long-term corrosion of cast irons and steel in marine and atmospheric environments", *Corrosion Science*, 68, 186–194.

Melchers, R. E. (2015), "Bi-modal trends in the long-term corrosion of copper and high copper alloys", *Corrosion Science*, 95, 51–61.

Melchers, R. E. and Jeffrey, R. (2008), "The critical involvement of anaerobic bacterial activity in modelling the corrosion behaviour of mild steel in marine environments", *Electrochimica Acta*, 54, (1), 80–85.

Melchers, R. E. and Li, C. Q. (2006), "Phenomenological modeling of reinforcement corrosion in marine environments", *ACI Materials Journal*, 103, (1), 25–32.

Melchers, R. E. and Li, C. Q. (2009a), "Long-term observations of reinforcement corrosion for concrete elements exposed to the North Sea", *Concrete in Australia,* 35, (2), 23–29.

Melchers, R. E. and Li, C. Q. (2009b), "Reinforcement corrosion in concrete exposed to the North Sea for over 60 years", *Journal of Science and Engineering: Corrosion,* 65, (8), 554–566.

Melchers, R. E. and Li, C. Q. (2009c), "Reinforcement corrosion initiation and activation times in concrete structures exposed to severe marine environments", *Cement and Concrete Research,* 35, 1068–1076.

Melchers, R. E., Li, C. Q. and Davision, M. A. (2009), "Observations and analysis of a 63 years old reinforced concrete promenade railing exposed to the North Sea", *Magazine of Concrete Research*, 61, (4), 233–243.

Melchers, R. E., Li, C. Q. and Lawanwisut, W. (2008), "Probabilistic modeling of structural deterioration of reinforced concrete beams under saline environment corrosion", *Structural Safety,* 30, (1), 447–460.

Meyers, M. A. and Chawla, K. K. (2008), *Mechanical Behavior of Materials,* Cambridge: Cambridge University Press.

Millard, S. G., Law, D., Bungey, J. H. and Cairns, J. (2001), "Environmental influences on linear polarisation corrosion rate measurement in reinforced concrete", *NDT and E International,* 34, (6), 409–417.

Mohebbi, H., Jesson, D. A., Mulheron, M. J. and Smith, P. A. (2010), "The fracture and fatigue properties of cast irons used for trunk mains in the water industry", *Materials Science and Engineering: A,* 527, (21–22), 5915–5923.

Mohebbi, H. and Li, C. Q. (2011), "Experimental investigation on corrosion of cast iron pipes", *International Journal of Corrosion,* 2011, 1–17.

Mok, D. H. B., Pick, R. J., Glover, A. G. and Hoff, R. (1991), "Bursting of line pipe with long external corrosion", *International Journal of Pressure Vessels and Piping,* 46, (2), 195–216.

Moore, T. J. and Hallmark, C. T. (1987), "Soil properties influencing corrosion of steel in Texas soils", *Soil Science Society of America Journal,* 51, (5), 1250–1256.

Muhammad, W., Dilan, R., Chun Qing, L. and Mojtaba, M. (2020), "Coupled effect of soil's acidity and saturation on pitting corrosion of buried cast iron", *Geotechnical Testing Journal,* 44, (3), 1–13.

Murray, J. N. and Moran, P. J. (1989), "Influence of moisture on corrosion of pipeline steel in soils using in situ impedance spectroscopy", *Corrosion,* 45, (1), 34–43.

Nakai, T., Matsushita, H. and Yamamoto, N. (2006), "Effect of pitting corrosion on the ultimate strength of steel plates subjected to in-plane compression and bending", *Journal of Marine Science and Technology,* 11, (1), 52–64.

Nastar, N. and Liu, R. (2019) *Failure Case Studies: Steel Structures.* Reston, VA: American Society of Civil Engineers.

National Water Commission Australia. (2010), *National Performance Report 2009–10: Urban Water Utilities.* Canberra: National Water Commission Australia.

National Water Commission Australia. (2013), *National Performance Report 2012–13.*

Nesic, S., Postlethwaite, J. and Olsen, S. (1996), "An electrochemical model for prediction of corrosion of mild steel in aqueous carbon dioxide solutions", *Corrosion,* 52, (4), 280–294.

Netto, T. A., Ferraz, U. S. and Estefen, S. F. (2005), "The effect of corrosion defects on the burst pressure of pipelines", *Journal of Constructional Steel Research,* 61, (8), 1185–1204.

Nguyen, K. T., Garbatov, Y. and Soares, C. G. (2013), "Spectral fatigue damage assessment of tanker deck structural detail subjected to time-dependent corrosion", *International Journal of Fatigue,* 48, 147–155.

Ni, Y., Ye, X. W. and Ko, J. M. (2010), "Monitoring-based fatigue reliability assessment of steel bridges: analytical model and application", *Journal of Structural Engineering,* 136, 1563–1573.

Nicholas, D. and Moore, G. (2009), "Corrosion of ferrous water mains: past performance and future prediction – a review", *Corrosion and Materials,* 34, (2), 33–40.

Nicholson, P. (1991), "Corrosion of municipal water systems", NACE *Corrosion 91 Conference,* Houston, TX, Paper No. 592.

Nie, X. H., Li, X. G., Du, C. W. and Cheng, Y. F. (2009a), "Temperature dependence of the electrochemical corrosion characteristics of carbon steel in a salty soil", *Journal of Applied Electrochemistry,* 39, (2), 277–282.

Nie, X., Li, X., Du, C., Huang, Y. and Du, H. (2009b), "Characterization of corrosion products formed on the surface of carbon steel by Raman spectroscopy", *Journal of Raman Spectroscopy,* 40, 76–79.

Noor, E. A. and Al-Moubaraki, A. H. (2008), "Corrosion behavior of mild steel in hydrochloric acid solutions", *International Journal of Electrochemical Science,* 3, (1), 806–818.

Norin, M. and Vinka, T. G. (2003), "Corrosion of carbon steel in filling material in an urban environment", *Materials and Corrosion*, 54, (9), 641–651.

Nürnberger, U. (2012), "Long-time behavior of non-galvanized and galvanized steels for geotechnical stabilization applications", *Materials and Corrosion*, 63, (12), 1173–1180.

Oriani, R. A. (1972), "A mechanistic theory of hydrogen embrittlement of steels", *Berichte der Bunsengesellschaft für physikalische Chemie*, 76, (8), 848–857.

Osarolube, E., Owate, I. O. and Oforka, N. C. (2008), "Corrosion behaviour of mild and high carbon steels in various acidic media", *Scientific Research and Essay*, 3, (6), 224–228.

Pantazopoulos, G. & Vazdirvanidis, A. (2013), "Cracking of underground welded steel pipes caused by HAZ sensitization", *Case Studies in Engineering Failure Analysis*, 1, (1), 43–47.

Pardo, A., Merino, M. C., Coy, A. E., Viejo, F., Arrabal, R. and Matykina, E. (2008), "Pitting corrosion behaviour of austenitic stainless steels–combining effects of Mn and Mo additions", *Corrosion Science*, 50, (6), 1796–1806.

Park, D. B., Lee, J. W., Lee, Y. S., Park, K. T. and Nam, W. J. (2009), "The effects of alloying elements on tensile strength and the occurrence of delamination in cold-drawn hyper-eutectoid steel wires", *Metals and Materials International*, 15, (2), 197–202.

Parkins, R. N. (1994), "The intergranular corrosion and stress corrosion cracking of mild steel in clarke's solution", *Corrosion science*, 36, (12), 2097–2110.

Payandeh, Y. and Soltanieh, M. (2007), "Oxide inclusions at different steps of steel production", *Journal of Iron and Steel Research, International*, 14, (5), 39–46.

Petersen, R. B., Dafter, M. and Melchers, R. E. (2013, November), Modelling the long-term corrosion of cast iron pipes, *Paper Presented at the 2013 Corrosion and Prevention Conference*, Brisbane, Australia.

Petersen, R. and Melchers, R. E. (2012, November), Long-term corrosion of cast iron cement lined pipes. *Paper Presented at the Annual Conference of the Australasian Corrosion Association 2012*, Melbourne, Australia.

Phull, B. (2003a), "Evaluating uniform corrosion", In S. D. Cramer and B. S. Covino (eds.), *Corrosion: Fundamentals, Testing, and Protection* (pp. 542–544), *ASM Handbook* (Vol. 13A), Materials Park, OH: ASM International.

Phull, B. (2003b), "Evaluating pitting corrosion", In S. D. Cramer and B. S. Covino (eds.), *Corrosion: Fundamentals, Testing, and Protection* (pp. 545–548), *ASM Handbook* (Vol. 13A), Materials Park, OH: ASM International.

Pritchard, O., Hallett, S. H., & Farewell, T. S. (2013), *Soil corrosivity in the UK—Impacts on critical infrastructure* (Working Paper Series), (pp. 1–55), London: Infrastructure Transitions Research Consortium.

Rajani, B. and Kleiner, Y. (2001), "Comprehensive review of structural deterioration of water mains: Physically based models", *Urban Water*, 3, (3), 151–164.

Rajani, B. and Makar, J. (2000), "A methodology to estimate remaining service life of grey cast iron water mains", *Canadian Journal of Civil Engineering*, 27, (6), 1259–1272.

Ralston, K. and Birbilis, N. (2010), "Effect of grain size on corrosion: A review", *Corrosion*, 66, (7), 075005–075005-13.

Raman, R. K. and Siew, W. H. (2014), "Stress corrosion cracking of an austenitic stainless steel in nitrite-containing chloride solutions", *Materials*, 7, (12), 7799–7808.

Ranji, A. R. and Zakeri, A. H. (2011), "Mechanical properties and corrosion resistance of normal strength and high strength steels in chloride solution", *Journal of Naval Architecture and Marine Engineering*, 7, (2), 94–100.

Rashidi, M. and Gibson, P. (2012), "A methodology for bridge condition evaluation", *Journal of Civil Engineering and Architecture*, 6, (9), 1149–1157.

Ren, R. K., Zhang, S., Pang, X. L. and Gao, K. W. (2012), "A novel observation of the interaction between the macroelastic stress and electrochemical corrosion of low carbon steel in 3.5 wt% NaCl solution", *Electrochimica Acta*, 85, 283–294.

Revie, R. and Uhlig, H. (2008), *Corrosion and Corrosion Control: An Introduction to Corrosion Science and Engineering*, Hoboken, NJ: John Wiley & Sons.

Roberge, P. R. (2007). *Corrosion Inspection and Monitoring* (Vol. 2). Hoboken, NJ: Wiley.

Rodrıguez, J. S., Hernández, F. S. and González, J. G. (2002), "XRD and SEM studies of the layer of corrosion products for carbon steel in various different environments in the province of Las Palmas (The Canary Islands, Spain)", *Corrosion Science*, 44, (11), 2425–2438.

Román, J., Vera, R., Bagnara, M., Carvajal, A. M. and Aperador, W. (2014), "Effect of chloride ions on the corrosion of galvanized steel embedded in concrete prepared with cements of different composition", *International Journal of Electrochemical Science*, 9, (2), 580–592.

Romanoff, M. (1957), *Underground Corrosion*, Washington, DC: US Government Printing Office.

Romanoff, M. (1964), Exterior corrosion of cast-iron pipe. *Journal American Water Works Association*, 56, (9), 1129–1143.

Romer, A. E. and Bell, G. E. (2001), "Causes of external corrosion on buried water mains", In J. P. Castronovo (ed.), *Pipelines 2001: Advances in Pipelines Engineering and Construction* (pp. 1–9), Reston, VA: American Society of Civil Engineers.

Rossum, J. R. (1969), Prediction of pitting rates in ferrous metals from soil parameters, *Journal American Water Works Association*, 61, (6), 305–310.

Sadiq, R., Rajani, B. and Kleiner, Y. (2004), "Probabilistic risk analysis of corrosion associated failures in cast iron water mains", *Reliability Engineering and System Safety*, 86, (1), 1–10.

Saha, J. K. (2012), *Corrosion of Constructional Steels in Marine and Industrial Environment: Frontier Work in Atmospheric Corrosion*, New York: Springer Science and Business Media.

Saheb, M., Neff, D., Bellot-Gurlet, L. and Dillmann, P. (2011), "Raman study of a deuterated iron hydroxycarbonate to assess long-term corrosion mechanisms in anoxic soils", *Journal of Raman Spectroscopy*, 42, (5), 1100–1108.

Sancy, M., Gourbeyre, Y., Sutter, E. M. M. and Tribollet, B. (2010), "Mechanism of corrosion of cast iron covered by aged corrosion products: Application of electrochemical impedance spectrometry", *Corrosion Science*, 52, (4), 1222–1227.

Sanders, P. F. and Hamilton, W. (1985), Biological and corrosion activities of sulphate-reducing bacteria in industrial process plant. In S. C. Dexter (ed.), *Biologically Induced Corrosion* (pp. 47–68), Houston, TX: National Association of Corrosion Engineers.

Schwerdtfeger, W. J. (1953), Laboratory measurement of the corrosion of ferrous metals in soils. *Journal of Research of the National Bureau of Standards*, 50, (6), 329–336.

Scott, G. N. (1934), Adjustment of soil corrosion pit depth measurements for size of sample. *Proceedings of the American Petroleum Institute*, 14, (4), 204.

Seica, M. V. and Packer, J. A. (2004), Mechanical properties and strength of aged cast iron water pipes. *Journal of Materials in Civil Engineering*, 16, (1), 69–77.

Shanmugam, S., Ramisetti, N. K., Misra, R. D. K., Mannering, T., Panda, D. and Jansto, S. (2007), "Effect of cooling rate on the microstructure and mechanical properties of Nb-microalloyed steels", *Materials Science and Engineering: A*, 460, 335–343.

Shimada, M., Kokawa, H., Wang, Z. J., Sato, Y. S. and Karibe, I. (2002), "Optimization of grain boundary character distribution for intergranular corrosion resistant 304 stainless steel by twin-induced grain boundary engineering", *Acta Materialia*, 50, (9), 2331–2341.

Shin, D. H., Kim, I., Kim, J. and Park, K. T. (2001), "Grain refinement mechanism during equal-channel angular pressing of a low-carbon steel", *Acta Materialia*, 49, (7), 1285–1292.

Shrestha, T., Alsagabi, S. F., Charit, I., Potirniche, G. P. and Glazoff, M. V. (2015), "Effect of heat treatment on microstructure and hardness of grade 91 steel", *Metals*, 5, (1), 131–149.

Silverman, D. C. (2003), "Aqueous corrosion", In S. D. Cramer and B. S. Covino (eds.), *Corrosion: Fundamentals, Testing, and Protection,* ASM Handbook (Vol. 13A), Materials Park, OH: ASM International.

Simonelli, G., Pasianot, R. and Savino, E. J. (1992), "Embedded-atom-method interatomic potentials for bcc-iron", *MRS Online Proceedings Library Archive*, 291, 567–572.

Sinyavskij, V., Ulanova, V. and Kalinin, V. (2004), "On the mechanism of intergranular corrosion of aluminum alloys", *Zashchita Metallov*, 40, (5), 537–546.

Soltis, J. (2015), "Passivity breakdown, pit initiation and propagation of pits in metallic materials–review", *Corrosion Science*, 90, 5–22.

Souza, R. D., Benjamin, A. C., Vieira, R. D., Freire, J. L. F. and Castro, J. T. P. (2007), "Part 4: Rupture tests of pipeline segments containing long real corrosion defects", *Experimental Techniques*, 31, (1), 46–51.

Stratmann, M and Müller, J., (1994), "The mechanism of the oxygen reduction on rust-covered metal substrates", *Corrosion Science*, 36, (2), 327–359.

Sun, F., Li, X. and Cheng, X. (2014), "Effects of carbon content and microstructure on corrosion property of new developed steels in acidic salt solutions", *Acta Metallurgica Sinica (English Letters)*, 27, (1), 115–123.

Suprunchuk, V. K., Shchegolev, V. M., Gratsianskii, N. N., Vdovenko, I. D. and Kolomiets, L. P. (1967), "A study of the effect of corrosion on mechanical properties of steels St. 2 and St. 3", *Materials Science*, 3, (4), 301–303.

Syugaev, A. V., Lomaeva, S. F., Reshetnikov, S. M., Shuravin, A. S., Sharafeeva, E. F. and Surnin, D. V. (2008), "The effect of the structure-phase state of iron-cementite nanocomposites on local activation processes", *Protection of Metals*, 44, (4), 367–371.

Szeliga, M. J. and Simpson, D. M. (2001), "Corrosion of ductile iron pipe: Case histories", *Materials Performance*, 40, (7), 22–26.

Szklarska-Smialowska, S. and Cragnolino, G. (1980), "Stress corrosion cracking of sensitized type 304 stainless steel in oxygenated pure water at elevated temperatures, *Corrosion*, 36, (12), 653–665.

Szklarska-Śmialowska, Z., Szummer, A. and Janik-Czachor, M. (1970), "Electron microprobe study of the effect of sulphide inclusions on the nucleation of corrosion pits in stainless steels", *British Corrosion Journal*, 5, (4), 159–161.

Thomas, B. G. (2001), Continuous casting of steel, In O. Yu (ed.), *Modeling for Casting and Solidification Processing* (pp. 499–540), New York: Marcel Dekker.

Tromans, D. (1994), "On surface energy and the hydrogen embrittlement of iron and steels", *Acta metallurgica et materialia*, 42, (6), 2043–2049.

Tsutsumi, Y., Nishikata, A. and Tsuru, T. (2007), "Pitting corrosion mechanism of Type 304 stainless steel under a droplet of chloride solutions", *Corrosion Science*, 49, (3), 1394–1407.

Turnbull, A. (2014), "Corrosion pitting and environmentally assisted small crack growth", *Proceedings of the Royal Society A: Mathematical, Physical and Engineering Sciences*, 470, (2169), 20140254.

Turnbull, A., McCartney, L. N. and Zhou, S. (2006), "A model to predict the evolution of pitting corrosion and the pit-to-crack transition incorporating statistically distributed input parameters", *Corrosion Science*, 48, 2084–2105.

Uhlig, H. H. (2011), *Uhlig's Corrosion Handbook* (vol. 51), New York: John Wiley and Sons.

Umemoto, M., Todaka, Y. & Tsuchiya, K. (2003), "Mechanical properties of cementite and fabrication of artificial pearlite", *Materials Science Forum*, 426, (432), 859–864.

Usher, K. M., Kaksonen, A. H., Cole, I. and Marney, D. (2014), "Critical review: Microbially influenced corrosion of buried carbon steel pipes", *International Biodeterioration and Biodegradation*, 93, 84–106.

Valiev, R. Z. and Langdon, T. G. (2006), "Principles of equal-channel angular pressing as a processing tool for grain refinement", *Progress in Materials Science*, 51, (7), 881–981.

Valor, A., Caleyo, F., Rivas, D. and Hallen, J. M. (2010), "Stochastic approach to pitting-corrosion-extreme modelling in low-carbon steel", *Corrosion Science*, 52, (3), 910–915.

Vertnik, R. and Šarler, B. (2014), "Solution of a continuous casting of steel benchmark test by a meshless method", *Engineering Analysis with Boundary Elements*, 45, 45–61.

Vodka, O. (2015), "Computation tool for assessing the probability characteristics of the stress state of the pipeline part defected by pitting corrosion", *Advances in Engineering Software*, 90, 159–168.

Wang, W.G. (2018), *Failure Analysis of Underground Pipeline Subjected to Corrosion* (Doctoral thesis), Melbourne: RMIT University.

Wang, W.G., Li, C. Q., Robert, D. and Zhou, A. (2018a), "Experimental investigation on corrosion effect on mechanical properties of buried cast iron pipes". *Journal of Materials in Civil Engineering*, 30, (8), 04018197.

Wang, W. G., Li, C. Q. and Shi, W. H. (2019a), "Degradation of mechanical property of corroded water pipes after long service", *Urban Water*, 16, (7), 494–504.

Wang, W. G., Robert, D., Zhou, A. N. and Li, C. Q., (2018b), "Factors affecting corrosion of buried cast iron pipes", *ASCE Journal of Materials in Civil Engineering*, 30, (11), 04018272.

Wang, W. G., Shi, W. H. and Li, C. Q. (2019b), "Time dependent reliability analysis for cast iron pipes subjected to pitting corrosion", *International Journal of Pressure Vessels and Piping*, 175, 103935.

Wang, X., Tang, X., Wang, L., Wang, C. and Guo, Z. (2014), "Corrosion behavior of X80 pipeline steel under coupling effect of stress and stray current", *International Journal of Electrochemical Science*, 9, (8), 4574–4588.

Wang, W., Yang, W., Li, C. Q. and Yang, S. (2020), "A new method to determine elastoplastic J-integral for steel pipes with longitudinal semi-elliptical surface cracks", *Engineering failure analysis*, 118, 104915.

Wang, W., Zhou, A., Fu, G., Li, C. Q., Robert, D. and Mahmoodian, M. (2017), "Evaluation of stress intensity factor for cast iron pipes with sharp corrosion pits", *Engineering Failure Analysis*, 81, 254–269.

Wasim, M. (2018), *External Corrosion and Its Effects on Mechanical Properties of Buried Metal Pipes* (Doctoral thesis), Melbourne: RMIT University.

Wasim, M., Li, C. Q., Mahmoodian, M. and Robert, D. (2019a), "Mechanical and microstructural evaluation of corrosion and hydrogen induced degradation of steel", *ASCE Journal of Materials in Civil Engineering*, 31, (1), 04018349.

Wasim, M., Li, C. Q., Mahmoodian, M. and Robert, D. (2019b), "Quantitative study of coupled effect of soil acidity and saturation on corrosion and microstructure of buried cast iron", *ASCE Journal of Materials in Civil Engineering*, 31, (9), 04019206.

Wasim, M., Li, C. Q., Robert, D. and Mahmoodian, M. (2020a), "Effect of soil's acidity and saturation on degradation of fracture toughness of buried cast iron", *ASCE Journal of Materials in Civil Engineering*, 32, (7), 04020180.

Wasim, M., Li, C. Q., Robert, D. and Mahmoodian, M. (2020b), "Fracture toughness degradation of cast iron due to corrosive mediums", *International Journal of Pressure Vessels and Piping*, 186, 104151.

Woodtli, J. and Kieselbach, R. (2000), "Damage due to hydrogen embrittlement and stress corrosion cracking", *Engineering Failure Analysis*, 7, (6), 427–450.

Wu, Y. H., Liu, T. M., Luo, S. X. and Sun, C. (2010), "Corrosion characteristics of Q235 steel in simulated Yingtan soil solutions", *Materialwissenschaft Und Werkstofftechnik*, 41, (3), 142–146.

Xi, X., Yang, S. T. and Li, C. Q. (2018a), "Accurate cover crack modelling of reinforced concrete structures subjected to non-uniform corrosion", *Structure and Infrastructure Engineering*, 14, (9), 1–13.

Xi, X., Yang, S. T. and Li, C. Q. (2018b), "A non-uniform corrosion model and meso-scale fracture modelling of concrete", *Cement and Concrete Research*, 108, 87–102.

Xi, X., Yang, S. T. and Li, C. Q., Cai, M., Hu, X. and Shipton, Z. (2018c), "Meso-scale mixed-mode fracture modelling of reinforced concrete structures subjected to non-uniform corrosion", *Engineering Fracture Mechanics*, 199, 114–130.

Xu, L. Y. and Cheng, Y. F. (2012), "An experimental investigation of corrosion of X100 pipeline steel under uniaxial elastic stress in a near-neutral pH solution", *Corrosion Science*, 59, 103–109.

Yamamoto, K., Mizoguti, S., Yoshimitsu, K. and Kawasaki, J. (1983), "Relation between graphitic corrosion and strength-degradation of cast iron pipe", *Corrosion Engineering*, 32, (3), 157–162.

Yamini, H. (2009), *Probability of Failure Analysis and Condition Assessment of Cast Iron Pipes Due to Internal and External Corrosion in Water Distribution Systems* (Doctoral thesis), University of British Colombia, Vancouver.

Yan, H. and Li, J. (2015), *Spatial Similarity Relations in Multi-scale Map Spaces*, Springer International Publishing, Switzerland.

Yang, S. (2010), *Concrete Crack Width under Combined Reinforcement Corrosion and Applied Load* (Doctoral thesis), University of Greenwich, London.

Yang, W., Fu, G. and Li, C. (2018), "Reliability-based service life prediction of corrosion-affected metal pipes with mixed mode fracture", *Journal of Engineering Mechanics*, 144, (2), 1–9.

Yang, S. T. and Li, C. Q. (2011), "Numerical prediction for corrosion-induced concrete crack width", *Proceedings of ICE: Construction Materials,* 164, (6), 293–303.

Yang, S. T., Li, K. F. and Li, C. Q. (2018b), "Analytical model for non-uniform corrosion-induced concrete cracking", *Magazine of Concrete Research*, 70, (1), 1–10.

Yang, S. T., Li, K. F. and Li, C. Q. (2018c), "Numerical determination of concrete crack width for corrosion-affected concrete structures", *Computers and Structures*, 207, 75–82.

Yang, S. T., Xi, X., Li, K. F. and Li, C. Q. (2018d), "Numerical Modelling of Non-Uniform Corrosion Induced Concrete Crack Width", *ASCE Journal of Structural Engineering*, 144, (8), 04018120.

Yi, M. Z. and Lin, J. S. (1990), "A study on diffusion coefficient of oxygen in steel", *Chinese Journal of Mechanical Engineering*, 3, (2), 1.

Yu, X., Chen, S., Liu, Y. and Ren, F. (2010), "A study of intergranular corrosion of austenitic stainless steel by electrochemical potentiodynamic reactivation, electron back-scattering diffraction and cellular automaton", *Corrosion Science*, 52, (6), 1939–1947.

Zhang, L. and Thomas, B. G. (2003), Inclusions in continuous casting of steel, *Paper Presented at the XXIV National Steelmaking Symposium*, Morelia, Mich, Mexico.

Zhao, Z., Haldar, A. and Breen Jr, F. L. (1994). "Fatigue-reliability evaluation of steel bridges", *Journal of Structural Engineering*, 120, 1608–1623.

Zhao, M. C., Yin, F., Hanamura, T., Nagai, K. and Atrens, A. (2007), "Relationship between yield strength and grain size for a bimodal structural ultrafine-grained ferrite/cementite steel", *Scripta materialia*, 57, (9), 857–860.

Zhou, Y. and Yan, F. (2016), "The relation between intergranular corrosion and electrochemical characteristic of carbon steel in carbonic", *International Journal of Electrochemical Science*, 11, 3976–3986.

Zucchi, F., Grassi, V., Monticelli, C. and Trabanelli, G. (2006), "Hydrogen embrittlement of duplex stainless steel under cathodic protection in acidic artificial sea water in the presence of sulphide ions", *Corrosion Science*, 48, (2), 522–530.

Index

acceleration of corrosion 25
  factor 173, 175, 177, 179, 184
  hydrochloric acid 25, 43, 91, 137, 154
  sulfuric acid 44, 91, 94
affecting factors 9
  defects 28, 33, 111, 149, 164
  microstructure 23, 27
  pH 26, 43, 50, 65, 79
  relative humidity 25, 26, 92
  salts 24, 26, 77, 80
  temperature 9, 23, 25, 46, 151

calibration 172
  acceleration factor 173, 176, 177, 179, 182
  scaling factor 172, 173, 175
cast iron 2
  brittle 78, 104, 105, 119, 167
  carbon content 2, 15, 27, 45, 124
  graphite 35, 36, 93, 105, 130
corrosion 17
  anodic reaction 17, 18
  electrochemical process 17, 21, 24
  cathodic reaction 18, 19, 25
corrosion current 20, 32, 45, 48, 96
  corrosion rate 9, 20, 37, 49, 97
  mass loss 20, 30, 32, 49, 57, 98
  pit depth 53, 63, 96, 103, 126
corrosion of ferrous metals 13
  cast iron corrosion 36, 90, 103, 117, 130
  ductile iron corrosion 36, 119, 120
  steel corrosion 13, 17, 19, 30, 36
corrosion measurement 9
corrosion tests 36
  burial test 36, 90, 93, 110
  immersion test 43, 45, 51, 94, 137
  natural corrosion 42, 46, 64, 95, 138
corrosion types 21
  crevice corrosion 23, 26
  microbial corrosion 21, 24
  pitting corrosion 22, 32, 38, 80, 132
  uniform corrosion 22, 31, 38, 104, 107

degradation 55
  reduction of failure strain 64, 65, 144
  reduction of fatigue strength 66–70
  reduction of fracture toughness 70, 75, 77, 110, 120
  reduction of modulus of rupture 107, 109, 110, 177, 187
  reduction of ultimate strength 60, 63, 66, 143, 181
  reduction of yield strength 55, 60, 87, 144, 182
delamination 34
  preferred corrosion 34, 135, 146, 152, 154
  quantification of delamination 53, 155
  solidification 35, 147–149, 152
ductile iron 35
  graphite 35, 36, 119, 124, 130
  nodular 35, 89, 119

effect of corrosion 30
  hydrogen concentration 60, 162–164, 169
  mass loss 30, 51, 57, 61, 144
  microstructural change 30, 33, 77, 131, 140
electrochemical process 17
  oxidation reaction 17, 18
  reduction reaction 18

fatigue strength 16
  number of cycles 16, 66–68, 71
  stress range 32, 66–69, 71
ferrous metals 1
  cast iron 35, 89, 104, 124, 133
  ductile iron 35, 119, 122, 129, 132
  steel 1, 13, 87, 151, 180

fracture toughness 45
single-edge notched bending (SENB) 45, 70, 71, 93, 120

hydrogen embrittlement 57
hydrogen concentration 57, 60, 162, 164, 167
mechanism 162, 165, 168, 169

mechanical properties 16
fatigue strength 16, 66, 67, 70, 183
fracture toughness 17, 70, 77, 110, 120
modulus of rupture 107, 110, 117, 177, 186
tensile strength 16, 75, 122, 185, 186
mechanisms of degradation 78
changes in element composition 78, 124
changes in grain size 84
changes in iron phase 85, 129
microstructure 20
element composition 33, 78, 124, 127, 143
grain size 27, 84, 141, 149, 158
iron phase 20, 85, 129, 142, 150
model of degradation 55
model of fatigue strength 66–70
model of fracture toughness 70, 75, 77, 110, 120
model of modulus of rupture 107, 109, 110, 177, 187
model of ultimate strength 60, 63, 66, 143, 181
model of yield strength 55, 60, 87, 144, 182

nanomechanics of corrosion 43
atomic lattice 77, 124, 194, 195, 198
bulk steel 43, 86, 199
interatomic potential 194, 195, 198
quantum mechanics 194

preferred corrosion 135
continuous casting 135, 151, 152
delamination 146, 153, 156, 158, 160
solidification 147, 148, 149, 156, 160

similarity theory 172
dynamic similarity 174
geometric similarity 172–174
kinematic similarity 173
scaling factor 172, 173, 175
simultaneous corrosion and loads 189
corrosion and bending 189, 190
corrosion and fatigue 191, 192
soil 28
moisture content 29, 92, 111
pH 29, 52, 73, 75, 91
resistivity 28, 29, 92
saturation 92, 99, 101, 115, 126
steel 10
atomic lattice 77, 124, 194, 195, 198
bulk steel 43, 86, 199
chemical composition 14, 23, 27, 35, 83
mechanical properties 10, 16, 41, 75, 143
strain 16
failure strain 16, 54, 64, 145, 165
stress effect 136
stress effect on corrosion 136, 137
stress effect on mechanical properties 143–146
stress effect on microstructure 140–143
stress-strain curve 56
failure strain 65, 144, 167
ultimate strength 61, 63, 144, 166
yield strength 56, 58, 59, 144, 166

tensile strength 55
modulus of rupture 107, 109, 110, 177, 187
ultimate strength 60, 63, 66, 143, 181
yield strength 55, 60, 87, 144, 182

Printed in the United States
by Baker & Taylor Publisher Services